In *Mathematician's Delight*, one of the first Pelicans to be published, W. W. Sawyer described the traditional mathematics of the engineer and scientist. In this new book the emphasis is not on those branches of mathematics which have great practical utility, but on those which are exciting in themselves: mathematics which is strange, novel, apparently impossible – for instance, an arithmetic in which no number is bigger than four. These topics are preceded by an analysis of the enjoyment of mathematics. Professor Sawyer not only shows what mathematicians get out of mathematics but also what they are trying to do, how mathematics grows, and why there are new mathematical discoveries still to be made. His aim is to give an all-round picture of the subject, and he therefore begins by describing the relationship between pleasure-giving mathematics and that which is the servant of technical and social advance.

For a complete list of books available please write to Penguin Books whose address can be found on the back of the title page

In *Mathematician's Delight*, one of the most popular Pelicans so far published, W. W. Sawyer described the traditional mathematics of the engineer and scientist. In this new book the emphasis is not on those branches of mathematics which have great practical utility, but on those which are exciting in themselves: mathematics which is strange, novel, apparently impossible, for instance, an arithmetic in which no number is bigger than four. These topics are preceded by an analysis of that enviable attribute 'the mathematical mind'. Professor Sawyer not only shows what mathematicians get out of mathematics, but also what they are trying to do, how mathematics grow, and why there are new mathematical discoveries still to be made. His aim is to give an all-round picture of his subject, and he therefore begins by describing the relationship between pleasure-giving mathematics and that which is the servant of technical and social advance.

PELICAN BOOKS

A 327

PRELUDE TO MATHEMATICS

W. W. SAWYER

Prelude to Mathematics

W. W. SAWYER

PENGUIN BOOKS

Penguin Books Ltd, Harmondsworth, Middlesex
U.S.A.: Penguin Books Inc., 3300 Clipper Mill Road, Baltimore 11, Md
CANADA: Penguin Books (Canada) Ltd, 47 Green Street,
Saint Lambert, Montreal, P.Q.
AUSTRALIA: Penguin Books Pty Ltd, 762 Whitehorse Road,
Mitcham, Victoria
SOUTH AFRICA: Penguin Books (S.A.) Pty Ltd, Gibraltar House,
Regent Road, Sea Point, Cape Town

—

First published 1955

Made and printed in Great Britain
by Western Printing Services Ltd
Bristol

CONTENTS

ACKNOWLEDGEMENTS

The quotation at the head of Chapter 9 is from Laurence Housman's *The Perfect One*, published by Jonathan Cape Ltd, by whose permission it is reprinted. The quotation at the head of Chapter 11 is from Ralph Hodgson's *Reason has Moons* and is used by permission of the author and Macmillan & Co. Ltd, the publishers.

CHAPTER ONE

On Beauty and Power

No mathematician can be a complete mathematician unless he is also something of a poet.

K. Weierstrass

Wisdom is rooted in watching with affection the way people grow.

Confucius

This is a book about how to grow mathematicians. Probably you have no intention of trying to grow mathematicians. Even so, I hope you may find something of interest here. I myself have no intention of growing plants. I never do any gardening if I can possibly get out of it. But I like very much to look at gardens other people have grown. And I am still more interested if I can meet a man who will explain to me (what very few gardeners seem able to do) just *how* a plant grows; how, when it is a seed under the earth, it knows which way is up for its stem to grow, and which way is down for the roots; how a flower manages to face towards the light; what chemical elements the plant needs from the soil, and just how it manages to rearrange them into its own living tissue. The interest of these things is quite independent of whether one actually intends to go out and do some hoeing.

What I am trying to do here is to write not from the viewpoint of the practical grower, but for the man who wants to understand what growth is. I am not writing for the professional teacher of mathematics (though teachers may be able to make practical applications of the ideas given here) but for the person who is interested in getting inside the mind of a mathematician.

It is very difficult to communicate the things that are really worth communicating. Suppose, for instance, that you have spent some years in a certain place, and that these years are particularly significant for you. They may have been years of early childhood, or school days, or a period of adult life when new experiences, pleasant or unpleasant, made life unusually interesting. If you revisit this place, you see it in a special way. Your companions, seeing it for the first time, see the physical scene, a pleasant village, a drab town street, whatever it may be. They do not see the essential thing that makes you want to visit the place; to make them see it, you need to be something of a poet; you have to speak of things, but so as to convey what you feel about those things.

But such communication is not impossible. Generally speaking, we overestimate the differences between people. I am sure that if one could go and actually be somebody else for a day, the change would be much less than one anticipated. The feelings would be the same, but hitched on to different objects. Most human misunderstandings are due to the fact that people talk about objects, and forget the varying significance the same object can have for different people.

Generally speaking, teaching conveys thoughts about objects rather than living processes of thought. Suppose someone comes to me with some kind of puzzle; it may be a question in a child's arithmetic book, or a serious problem of scientific research. Perhaps I succeed in solving the puzzle. Then it is quite easy to explain the solution. Suppose I do so; I have shown the questioner how to deal with that particular problem. But if another problem, of a different kind, arises, I shall be consulted again. I have not made my pupil independent of me. What would be really satisfactory would be if I could convey, not simply the knowledge of how to solve a particular puzzle, but the living attitude of mind that would enable my pupil to attack puzzles successfully without help from anyone.

Naturally, there must be certain limitations to what one can expect. Intelligence is one of the factors in problem solving, and it may well be inborn. But there are many other factors – emotions of fear or confidence, habits of self-reliance, initiative, persistence – which depend on education. I do not believe our ancestors at the time of the cave men differed at all from us in inborn qualities of intellect. All historical changes from that time to this, all differences in institutions between one country and another, have essentially been changes in education.[1]

Pelican Books are themselves a symptom of a profound his-

1. Kluckhohn in *Mirror for Man* gives an interesting example of an American boy brought up in China. In physical appearance the boy was American; in all mental and emotional qualities, Chinese.

torical change. That there should be in so many different countries of the world a large body of thoughtful men and women reading, studying, forming a kind of invisible, international university – a century ago, or in any previous age one would have looked for such a thing in vain.

At the present time, when our knowledge of material things is so great, and our understanding of ourselves so small, a true appreciation of the enormous unused potentialities of education is essential. The industrial revolution implies and requires a psychological revolution. Psychologically, we still belong to the era when people refused to believe that locomotives would run.

The present book was worked out in a country where a great educational change was taking place. In 1948, the University College of the Gold Coast was founded. The students were keen and of first-rate ability. So far as inborn intelligence went, they were capable of becoming, within ten or twenty years, research workers in mathematics, university lecturers, professors. But of course there was no mathematical tradition in the country. That had to be created.

It was therefore necessary to obtain a sort of essence of mathematics; to examine the life of a budding mathematician in one of the older mathematical centres; to study all the influences that helped him to grow; the atmosphere of school and college, countless hints, allusions, suggestions from older mathematicians and from books. From all of this to try to form some clear idea of what we were trying to do, what the qualities of a mathematician were, how they were to be stimulated.

In the first five chapters of this book, I try to give a specification of what a mathematician is, and how he grows. These chapters also contain various pieces of mathematics to illustrate what interests a mathematician. The remainder of the book is an exposition of various branches of mathematics; these have been selected for their strangeness, their novelty, their stimulating power. Moreover, they are elementary. A confused recollection of School Certificate mathematics should be sufficient to see a reader through them.

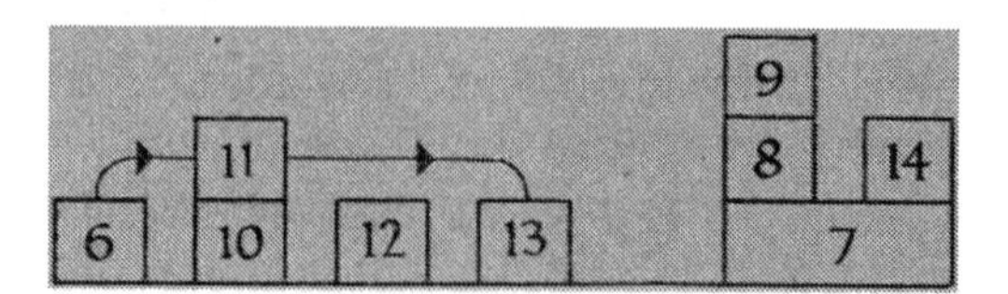

Figure 1 – THE STRUCTURE OF THIS BOOK

Chapters 1–5 use mathematics to illustrate the qualities of a mathematician. They are not shown in the diagram.

Where one chapter is shown as resting on another, the upper chapter makes use of the ideas explained in the lower one. Thus, it is necessary to read Chapter 10 before Chapter 11.

An arrow going from one chapter to another indicates some connexion between the two chapters. For instance, one section of Chapter 13 (the section 'Finite Geometries') cannot be understood without reference to Chapter 11. But all the rest of Chapter 13 could be read as the first chapter of the book.

It will be seen that bungalows rather than skyscrapers predominate.

Calculus is referred to once or twice, but is nowhere used as part of the argument. This fact is interesting, as showing that some parts of recent mathematics (i.e. since 1800) are not a development of the older work, but have gone off in quite a new direction.

Nor will you find much in the way of long calculations here. Nearly every mathematical discovery depends upon a fairly simple idea. Textbooks often conceal this fact. They contain massive calculations, and convey the impression that mathematicians are men who sit at desks and use up vast quantities of stationery. This impression is quite wrong. Many mathematicians can work quite happily in a bath, in bed, while waiting at a railway station, or while cycling (preferably not in traffic). The calculations are made before or after. The discovery itself grows from a central idea. It is these central ideas I hope to convey. Naturally, some details must be given, if the book is to be mathematics at all, and not merely a sentimental rhapsody on the mathematical life.

But before you part with your money, I want to warn you that there will be one or two pages on which you may see some lengthy algebraic expression. One of the things I want to discuss is *how to look at an algebraic expression*. The longer the expression, the more important it is to know how to look at it. For instance, in Chapter 9, on determinants, you will see some long, sprawling chains of symbols. These are only introduced so that we can exclaim, 'Look at this mess! How are we to discover, in this apparent chaos, some simple idea that we can think about? Where lies the shape, the form, the pattern, without which nothing can be regarded as mathematics?'

A point that should be borne in mind is that, generally speaking, higher mathematics is simpler than elementary mathematics. To explore a thicket on foot is a troublesome business; from an aeroplane the task is easier.

One thing most emphatically this book does not claim to do; that is, to give a balanced account of the development of mathematics since 1800, nor even to give a rounded account of what mathematicians are doing to-day. This book presents samples of mathematics. It does not claim to do more. Indeed, it is doubtful if more could be done in a book of this size.

THE EXTENT OF MATHEMATICS

Very few people realize the enormous bulk of contemporary mathematics. Probably it would be easier to learn all the languages of the world than to master all mathematics at present known. The languages could, I imagine, be learnt in a lifetime; mathematics certainly could not. Nor is the subject static. Every year new discoveries are published. In 1951 merely to print *brief summaries* of a year's mathematical publications required nearly 900 large pages of print. In January alone the summaries had to deal with 451 new books and articles. The publications here mentioned dealt with new topics; they were not restatements of existing knowledge, or very few of them were. To keep pace with the growth of mathematics, one would have to read about fifteen papers a day, most of them packed with technical details and of considerable length. No one dreams of attempting this task.

The new discoveries that mathematicians are making are very varied in type, so varied indeed that it has been proposed (in despair) to define mathematics as 'what mathematicians do'. Only such a broad definition, it was felt, would cover all the things that might become embodied in mathematics; for mathematicians to-day attack many problems not regarded as mathematical in the past, and what they will do in the future there is no saying.

A little more precise would be the definition 'Mathematics is the classification of all possible problems, and the means appropriate to their solution'. This definition is somewhat too wide. It would include such things as the newspaper's 'Send your love problems to Aunt Minnie', which we do not really wish to include.

For the purposes of this book we may say, 'Mathematics is the classification and study of all possible patterns'. Pattern is here used in a way that not everybody may agree with. It is to be understood in a very wide sense, to cover almost *any kind of regularity that can be recognized by the mind.* Life, and certainly intellectual life, is only possible because there are certain regularities in the world.[1] A bird recognizes the black and yellow bands of a wasp; man recognizes that the growth of a plant follows the sowing of seed. In each case, a mind is aware of pattern.

Pattern is the only relatively stable thing in a changing world. To-day is never exactly like yesterday. We never see a face twice from exactly the same angle. Recognition is possible not because experience ever repeats itself, but because in all the flux of life certain patterns remain identifiable. Such an enduring pattern is implied when we speak of 'my bicycle' or 'the river Thames', notwithstanding the facts that the bicycle is rapidly rusting away and the river perpetually emptying itself into the sea.

Any theory of mathematics must account both for the power of mathematics, its numerous applications to natural science, and the beauty of mathematics, the fascination it has for the mind. Our definition seems to do both. All science depends on regularities in nature; the classification of types of regularity, of patterns, should then be of practical value. And the mind should find pleasure in such a study. In nature, necessity and desire are always linked. If response to pattern is characteristic alike of animal and human life, we should expect to find pleasure associated with the response to pattern as it is with hunger or sex.

It is interesting to note that pure mathematicians, moved only by their sense of mathematical form, have often arrived at ideas later of the utmost importance to scientists. The Greeks studied the ellipse more than a millennium before Kepler used their ideas to predict planetary motions. The mathematical theory needed by relativity was in existence thirty to fifty years before Einstein

1. Compare Poincaré, *Science and Method.*

found a physical application for it. Many other examples could be given.

On the other hand, many very beautiful theories, to which any pure mathematician would concede a place in mathematics for their intrinsic interest, arose first of all in connexion with physics.

NATURE'S FAVOURITE PATTERN?

Another striking fact is that in nature we sometimes find the same pattern again and again in different contexts, as if the supply of suitable patterns were extremely limited. The pattern which mathematicians denote by $\Delta^2 V$ occurs in at least a dozen different branches of science. It arises in connexion with gravitation, light, sound, heat, magnetism, electrostatics, electric currents, electromagnetic radiation, waves at sea, the flight of aeroplanes, vibrations of elastic bodies and the mechanics of the atom – not to mention a pure mathematical theory of first-class importance, the theory of functions $f(x + iy)$, where i is $\sqrt{-1}$.

Practical men often make the mistake of treating all these applications as quite separate and distinct. This is a great waste of effort. We have not twelve theories, but one theory with twelve applications. The same pattern appears throughout. Physically the applications are distinct, mathematically they are identical.

The idea of the same pattern arising in different circumstances is a simple one. One only has to invent a Greek name for this idea to have one of the commonest terms of modern mathematics – *isomorphic* (*isos*, like; *morphe*, shape – having the same shape). Nothing delights a mathematician more than to discover that two things, previously regarded as entirely distinct, are mathematically identical. 'Mathematics', said Poincaré, 'is the art of giving the same name to different things.'

One might ask, 'Why does this one pattern, $\Delta^2 V$, occur again and again?' Here we tremble on the brink of mathematical mysticism. There can be no final answer. For suppose we show that this pattern has certain properties which make it particularly suitable; we then have to ask, 'But why does Nature prefer those properties?' – in endless mazes lost. Nevertheless a certain answer can be given as to why $\Delta^2 V$ so frequently occurs.[1]

1. The explanation is along the following lines. In empty space every point is as good as every other point, and every direction as good as every

The impossibility of any final answer to the question, 'Why is the universe like it is?' does not therefore mean that the enquiry is entirely useless. We may succeed in discovering that all scientific laws so far discovered have certain properties in common. A mathematician, in studying what patterns have these properties, has a reasonable belief that his work will be useful to future generations – not a certainty, of course; nothing is certain. He may also hope to satisfy his own desire to achieve a deep insight into the workings of the universe.

THE MATHEMATICIAN AS CONSULTANT

Technical men and engineers do not, as a rule, have the vision of mathematics as a way of classifying *all* problems. They tend to learn those parts of mathematics which have been useful for their profession *in the past*. In the face of a new problem they are accordingly helpless. It is then that the mathematician gets called in. (This division of labour between engineers and mathematicians is probably justified; life is too short for the simultaneous study of technical practice and abstract pattern.) The encounter between the mathematician and the technician is usually amusing. The practical man, by daily contact with his machinery, is so soaked in familiarity with it that he cannot realize what it feels like to see the machine for the first time. He pours out a flood of details which, to the consulting mathematician, mean precisely nothing. After some time, the mathematician convinces the technical man that mathematicians are really ignorant, and that the simplest things have to be explained, as to a child or to Socrates. Once the mathematician understands what the machine does, or is required to do, he can usually translate the problem into mathematical terms. Then he can tell the practical men one of three things, (i) that the problem is a well-known one, already solved, (ii) that it is a new problem which perhaps he can do something about, (iii) that it is an old problem which mathematicians have tried to solve without success, and

other direction. Laws holding in empty space may therefore be expected not to single out any particular point or direction. This considerably restricts the choice of possible laws. $\Delta^2 V = 0$ expresses in symbols the law that the value of V at any point equals the average value of V on a sphere with centre at this point. This law treats all points and all directions alike, and is the simplest law that does so.

that several centuries may elapse before any advance is made with it; the factory must deal with the problem empirically. Situation (iii) occurs with distressing frequency. But situations (i) and (ii) do also occur, and it is then that the mathematician, by his interest in and classification of patterns, can be of service to trades and professions about which, in one sense, he knows nothing.

A mathematician who is interested in consultative work therefore needs not merely to study problems which have occurred, he must be prepared for those that may occur. To a certain extent, one can recognize that practical problems form a *type*. For example, very often a practical problem takes the form of a differential equation.[1] Some differential equations we know how to solve, others we do not. A mathematician may therefore seek to enlarge his armoury by studying those differential equations which have so far defied solution. And this will lead him to ask all sorts of fundamental questions, such as, 'What is the difference between the equations that have been solved and those that have not? What makes an equation easy or difficult to solve?'

But practical problems do not always conform to type. Sometimes problems arise which are quite unlike those of normal routine. The clue to their solution may be found quite by chance; perhaps they resemble some puzzle solved in an idle moment. When this happens, the puzzle may prove to be the foundation of a new and dignified mathematical theory. This doctrine, like all doctrines, is capable of abuse. A man may waste his life on footling little puzzles, and defend this on the ground that these *might* be the beginnings of new branches of mathematics. So they might: the matter depends on one's judgement of what is likely to prove important, and there is no rule by which to settle the question. And any mathematician will agree that there are subjects which have not yet found any technical application, but which one *feels* to be major parts of mathematics. They are part of the battle, they are not escapism. Some time in the future like the ellipse they will find their Kepler, like tensor analysis their Einstein. But in any case there they are all ready, massive machines to solve a certain class of problem, if the necessity arises.

1. See *M.D.*, Chapter 12, final section. The abbreviation *M.D.* will be used for references to *Mathematician's Delight*.

THE MATHEMATICIAN AS ARTIST

As I write this, I imagine a pure mathematician reading it with increasing distress; that is, supposing he gets as far as this. 'You are treating mathematics,' he will be saying, 'as something useful. But mathematics is not a means to something else; it is an end in itself. Not the usefulness of mathematics is important but the beauty. Technical mathematics is the dullest part of mathematics. Look at the people working on the theory of numbers, which has no application at all; would you prefer them to be working at bookkeeping?'

This view – which is sincerely held by many eminent and some great mathematicians – may be contrasted with its opposite, the utilitarian, bureaucratic view of mathematics. According to this somewhat Puritanical theory, mathematicians should be ashamed of their interest in beauty and elegance; they should only work at mathematics when some official summons them to solve an immediately useful problem.

Both views are incomplete. Either of them, pursued to a logical conclusion, would be fatal to mathematical and even to technical progress.

Let us examine the 'mathematics for mathematics' sake' theory first and consider (if I may take a particular example) its application to the Gold Coast. The Gold Coast is a country of lively, intelligent, and cheerful people; let nothing here said suggest that it is a miserable place. But it is a country with certain acute material needs. In many places water supply is uncertain, sanitation lacking, disease widespread, food inadequate. Some children grow up permanently hungry. How then are we to defend the expense of the mathematics department in the new university? On the grounds of the beauty of mathematics? To defend mathematics in such circumstances *purely* on the grounds of its beauty is the height of heartlessness. Mathematics has cultural value; but culture does not consist in stimulating oneself with novel patterns in indifference to one's surroundings. Obviously the power of mathematics, mathematics as a means to engineering and medical science, is of first-rate importance in any developing country; the beauty without the power is futile.

But there is still a word to be said for the artist. Power without beauty is liable to be impotent. An activity engaged in purely for its consequences, without any pleasure in the activity itself, is likely to be poorly executed. An engineer may begin to study mathematics because it is useful for his profession, but if his concern stops there, if he does not begin to feel the fascination of the subject itself, he will not do much good with mathematics.

The utilitarian view of mathematics is realistic in one way; it recognizes the fact that a mathematician is a human being, dependent on the efforts of other human beings for his food and clothing and shelter and fuel, and owing some return for these. This is the aspect of the matter most easily appreciated by non-mathematicians, by administrators and by taxpayers. Considerations of utility may show that a country needs mathematicians; but yet give no clue as to how it is to get them. It might be very desirable for the Sahara to be covered by a forest of oak trees, to give shade to travellers, but that does not cause trees to grow there. Mathematicians, like trees, are living organisms and will only grow in conditions where they can grow.

It might be objected that men are not trees; that if a man realizes something ought to be done, he can go and do it. This is true within certain limits. There can be social conditions favourable to mathematical studies; if a country urgently needs mathematicians, and if everyone knows this, mathematics may well flourish. But this still does not answer the question of *how* it comes to flourish. An external motive, good or bad, is not enough. Greed for money, desire for fame, love of humanity are equally incapable of making a man a composer of great music. It has been said that most young men would like to be able to sit down at the piano and improvise sonatas before admiring crowds. But few do it; to desire the end does not provide the means; to make music you must be interested in music, as well as (or instead of) in being admired. And to make mathematics you must be interested in mathematics. The fascination of pattern and the logical classification of pattern must have taken hold of you. It need not be the only emotion in your mind; you may pursue other aims, respond to other duties; but if it is not there, you will contribute nothing to mathematics.

To this extent, the artist is more realistic than the bureaucrat. The artist does at least understand how people become mathematicians. Both the pure artist and the pure bureaucrat are wrong, or at least incomplete. If the teaching of mathematics had to be based on the theories of either one of them, that of the artist would do less harm. An artist may be an anarchist, a bohemian, or a tramp, but at least he is alive, and without life there is no growth. If such a man can teach children to love a subject for itself, there is always the hope that at a later age these children will turn their gifts to useful ends. But if they are left in the hands of the extreme utilitarians, they will have no gifts to turn.

CHAPTER TWO

What are the Qualities of a Mathematician?

> I do not fancy this acquiescence in second-hand hearsay knowledge; for, though we may be learned by the help of another's knowledge, we can never be wise but by our own wisdom.
>
> Montaigne, *Of Pedantry*

Mental venturesomeness is characteristic of all mathematicians. A mathematician does not want to be told something; he wants to find it out for himself. An adult mathematician, of course, if he hears that some great discovery has been made will want to know what it is, he will not want to waste time re-discovering it. But I am thinking of mathematicians at an early age, where this mental aggressiveness is very marked. For example, if you are teaching geometry to a class of boys nine or ten years old, and you tell them that no one has ever trisected an angle by means of ruler and compasses alone, you will find that one or two boys will stay behind afterwards and attempt to find a solution. The fact that in two thousand years no one has solved this problem does not prevent them feeling that they might get it out during the dinner hour. This is not exactly a humble attitude, but neither does it necessarily indicate conceit. It is simply the readiness to respond to any challenge. In fact, of course, the trisection of the angle *by the means specified* has been proved to be impossible; it is in the same category as trying to express $\sqrt{2}$ as a rational fraction p/q.

Again, a good pupil will always be running ahead of the course. If you show him how to solve a quadratic equation by completing the square, he will want to know whether it is possible to solve a cubic equation by completing the cube. The rest of the class do not ask such a question. Having to solve quadratics is bad enough for them; they do not wish to add to their burdens.

The desire to explore thus marks out the mathematician. This is one of the forces making for the growth of mathematics. The mathematician enjoys what he already knows; he is eager for new knowledge. Fractional indices, in school algebra, seem to illustrate this point. I can well imagine anyone, after reading an elementary account of fractional and negative indices, wondering whether such things are really justified at all. There are many logical difficulties to be overcome. The discoverer of fractional indices, I feel, must have enjoyed working with ordinary indices and been so anxious to extend this subject that he was willing to take the logical risks. A new discovery is nearly always a matter of faith in the first instance; later, of course, when one has seen that it does work, one has to find a logical justification that will satisfy the most cautious critics.

Interest in pattern has already been mentioned. Pattern appears already in the first steps of arithmetic; in the fact, say, that four stones can be arranged to form a square, while five cannot. Mathematical, like musical, ability is apparent at a very early age; four years old, or even earlier. A young child once said to me, 'I like the word September. It goes sEptEmbEr'. I had never myself noticed this pattern .*..*..*. in the vowels and consonants of 'September'. It is of course perfectly symmetrical. Such a child should enjoy arithmetic.

A very elementary example of pattern is contained in the multiplication tables. Children usually like the 2 × and 5 × tables, because the final digits are easy to remember – always even in the 2 ×, and still better, always 0 or 5 in the 5 ×. But even the 7 × table has its regularities. If one examines the final digits of 7, 14, 21, 28, 35, 42, 49, 56, 63, 70 they are

$$7 \quad 4 \quad 1 \quad 8 \quad 5 \quad 2 \quad 9 \quad 6 \quad 3 \quad 0$$

with the differences -3 -3 $+7$; -3 -3 $+7$; -3 -3 -3, in which quite a definite rhythm is apparent.[1] The final digits of the 7 × table read backwards are of course those of the 3 × table. Even in the lowest forms at school, the habit of observing mathematical regularities can grow. Much of the early work of Gauss springs from his habit of making calculations and observing the results. Hermite, a great French mathematician, also

1. A very similar pattern occurs in the sharps and flats of key signatures.

The signature for seven sharps is 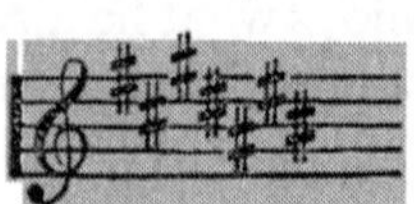

which gives the sequence of steps, down 3, up 4, down 3, down 3, up 4, down 3.

stressed the importance of observation in leading to mathematical discoveries.[1] Of course observation alone is not sufficient to make a great mathematician.

Just as a pupil who notices arithmetical regularities will find this helpful in doing arithmetic, so the observation of regularities in algebra helps one to avoid or detect slips.

For example, the condition for the quadratic equation $ax^2 + 2bx + c = 0$ to have equal roots is $b^2 - ac = 0$. (The reader may be more familiar with this in the form $b^2 - 4ac = 0$, which holds for $ax^2 + bx + c = 0$.) One can find a similar condition in connexion with a cubic equation. A cubic equation has three roots; we enquire in what circumstances two of these three roots are equal. For the equation $ax^3 + 3bx^2 + 3cx + d = 0$ the condition is $(bc-ad)^2 - 4(ac-b^2)(bd-c^2) = 0$. If you like, you can multiply out completely and write the condition as

$$a^2d^2 - 6abcd + 4b^3d + 4ac^3 - 3b^2c^2 = 0.$$

If you work with such expressions you can hardly help noticing certain things.

(i) The total number of letters in each term of the expression is the same. For instance, in $b^2 - ac$, each term contains *two* letters multiplied together, i.e. every term is of the second degree. In the longer condition for the cubic to have equal roots, every term contains *four* letters multiplied together. b^3d is of course short for *bbbd*. Every term is of the fourth degree.

(ii) It is not so obvious, but there is another kind of balance in these expressions, a balance between letters which occur early in the alphabet and those which occur late. For instance, in the condition for the cubic *abcd* occurs. The term a^2d^2, or *aadd*, also occurs. In the term *aadd*, the second letter *a* occurs earlier in the alphabet than *b*, the second letter of *abcd*. But justice is done. If we look at the third letters, *aadd* contains *d* while *abcd* contains the earlier letter *c*. The balance is exact; *b* is immediately after *a*, *c* is immediately before *d*. The same balance holds throughout. This may be checked in the following way. Suppose we let each term score in the following way. *a*, being the earliest letter in the alphabet, scores 0; *b* scores 1, *c* scores 2, *d* scores 3. It will be found that each term scores a total of 6. For example, *abcd* scores $0 + 1 + 2 + 3 = 6$; ac^3 scores $0 + 2 + 2 + 2$.

1. Hermite, *Collected Works*, Vol. IV, p. 586.

The technical term for this score is *weight*: we say that each term has the weight 6. In $b^2 - ac$, for a quadratic, each term has weight 2.

(iii) The sum of the coefficients in each expression is zero. In $b^2 - ac$ the coefficients are 1 and -1, which add up to 0. In $a^2d^2 - 6abcd + 4b^3d + 4ac^3 - 3b^2c^2$ the coefficients are 1, -6, 4, 4, -3 with the sum 0. Another way of expressing the same thing is to say that if we put a, b, c, d all equal to 1, the expression takes the value zero.

These three things we have noticed provide a check on our accuracy. (i) enables us to check if we have made a slip in writing the power to which any letter occurs, for such a slip will make the terms of unequal degree (unless of course we make several slips, which compensate each other in this respect). (ii) will save us from slips in copying the letters. If, for example, we copied d as a at some stage of the working, this would alter the weight of a term. Test (iii) will save us from making slips in adding the terms together. Suppose for example that we overlook a term, when we are collecting the terms together. On putting the value 1 for every letter we shall find we do not get zero, and our attention will be drawn to the mistake.

This type of checking does not give a profound insight into mathematics; it is one of the things a mathematician does almost unconsciously, to look at an answer and make sure it has the kind of symmetry, the kind of balance he expects. It is surprising how often pupils are not taught to examine things in some such way. In examinations, one again and again finds pupils handing in answers which – in the phrase of G. K. Chesterton's Father Brown – have 'the wrong shape', and which obviously are crying out for some simple correction. The principle extends far beyond the special type of work discussed above.

A GEOMETRICAL PATTERN

In school geometry one meets the result

$$x^2 = \frac{(ac + bd)(ad + bc)}{ab + cd}$$

where x, a, b, c, d stand for the lengths of the lines shown in Figure 2.

This result is very easily proved, by finding $\cos Q$ and $\cos S$ in terms of the sides of the triangles PQR and PSR. The sum of the two expressions must be zero, for $Q + S = 180°$ (cyclic quadrilateral), so $\cos S = -\cos Q$. The resulting equation is then solved for x^2.

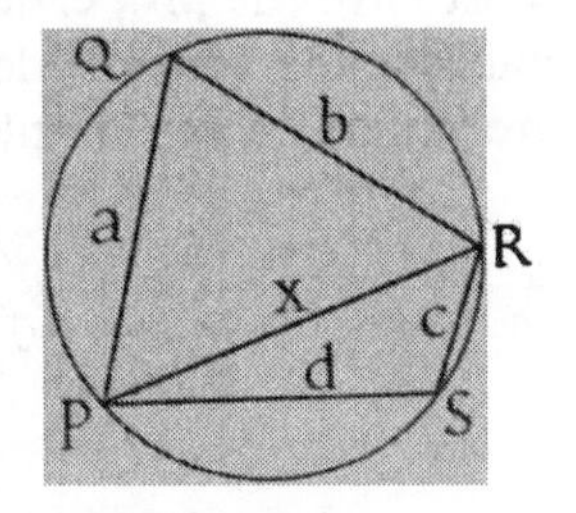

Figure 2

There is nothing very striking in this proof, but the pattern of the result is remarkable.

There are three ways of dividing four objects up into pairs. If four men play bridge, *A* and *B* are partners against *C* and *D*; or *A* and *C* against *B* and *D*; or *A* and *D* against *B* and *C*. No other arrangement is possible.

The algebraic expressions $ab + cd$, $ac + bd$, $ad + bc$ are also built by dividing up the four symbols *a, b, c, d* into pairs, and putting a + sign in the middle. These three, and only these three, can be formed in this way; $cd + ab$ of course is the same as $ab + cd$.

In the formula above all three expressions occur, two of them in the numerator, the remaining one in the denominator.

I do not here want to go into the question why this occurs, but simply to point out that this formula is, in a quiet way, a very memorable one.

SIGNIFICANCE AND GENERALIZATION

In other arts, if we see a pattern we can admire its beauty; we may feel that it has significant form, but we cannot say what the significance is. And it is much better not to try. A poet protested against the barbarous habit of requiring children to paraphrase poems. The only way you can explain the meaning of a poem, he said, is by writing a better poem, and that is a lot to ask of children.

But in mathematics it is not so. In mathematics, if a pattern occurs, we can go on to ask, Why does it occur? What does it signify? And we can find answers to these questions. In fact, for every pattern that appears, a mathematician feels he ought to know why it appears.

For example, we can explain why conditions for equal roots

must have the properties (i), (ii), and (iii) described earlier. I will indicate the general idea, and omit the algebraic details (which are actually quite simple).

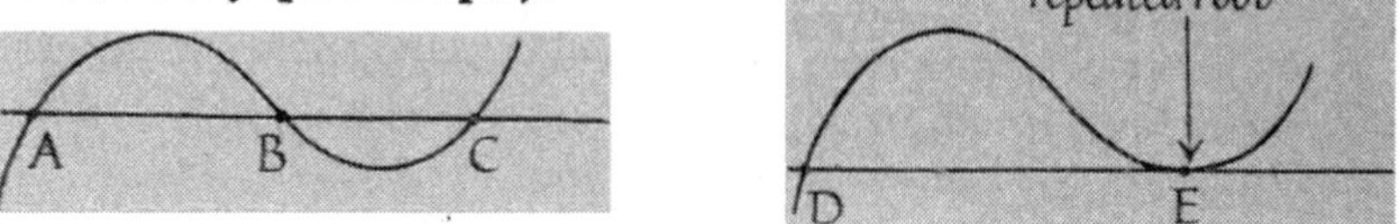

Figure 3a and b

Let us consider the graph of a cubic, which for short I will call $y = f(x)$. The first graph shown (Figure 3a) represents a cubic with all its roots real and unequal. But if the cubic were lifted up until the points B and C coincided, we should obtain the second graph, which represents a cubic with a pair of equal roots (Figure 3b). In passing it may be mentioned that calculus gives us a simple test for equal roots. At the point E, both

$$y \text{ and } \frac{dy}{dx}$$

are zero. Equal roots occur if the equation $f'(x) = 0$ has a root in common with the original equation $f(x) = 0$.

Now imagine that the second graph (Figure 3b) were drawn on a sheet of rubber. Suppose this rubber were to be stretched vertically; this would have the effect of changing the scale on the y-axis. The graph would become that of $y = kf(x)$. Now obviously if the original graph touched the x-axis (which it does when there are equal roots), the new graph, obtained by stretching, would still do so. That is to say, if $f(x) = 0$ has equal roots, so has $kf(x) = 0$. By considering the effect of the extra factor k on the constants a, b, c, d, the coefficients in $f(x)$, one can show that the condition for equal roots must have all its terms of equal degree – property (i).

In the same way, stretching horizontally will not affect the general appearance of the graph. It will still touch the x-axis. From this we may conclude that if $f(x) = 0$ has equal roots, so has $f(kx) = 0$. This leads us to property (ii), that all the terms have equal weight.

Property (iii) is the simplest of all. If we replace all the letters a, b, c, d by 1, the quadratic equation becomes $x^2 + 2x + 1 = 0$ and the cubic $x^3 + 3x^2 + 3x + 1 = 0$. These equations are $(x + 1)^2 = 0$ and $(x + 1)^3 = 0$, which obviously have the root -1

repeated. (All three roots of the cubic are -1; it would have been sufficient for our purpose if two of them had been -1.)

So properties (i) *and* (ii) *hold for any condition that is unaffected by a change of scale of y (a vertical stretch) and a change of scale of x (a horizontal stretch); property* (iii) *holds for any condition that is satisfied by* $(x + 1)^n = 0$.

This represents a great advance on the time when we had simply *noticed* the properties (i), (ii), (iii). We have now interpreted them; we know *why* they hold and *when* they hold.

And this allows us to apply tests (i), (ii), (iii) to types of condition other than those for equal roots. For example, with a cubic equation, which has three roots, we may ask in what circumstances one root occurs exactly mid-way between the other two. In the first cubic graph, drawn earlier, this would mean the point B is to be half-way from A to C. This property would not be destroyed by changes of scale, vertical or horizontal. This property holds for the equation $(x + 1)^3 = 0$; for this equation, the points A, B, C all coincide at $x = -1$, and B thus is the mid-point of AC. The algebraic condition for this property must therefore have the features (i), (ii), (iii). The condition is in fact $2b^3 - 3abc + a^2d = 0$, and as you can see it is of the form expected.

This is an example of *generalization*, one of the most important factors in the development of mathematics. We began with an observation which applied only to conditions for equal roots; we ended with a principle which applied to a much wider class of conditions. This is obviously valuable; the wider the situations to which a principle is relevant, the more often it is likely to help us out of our difficulties. As Poincaré said, 'Suppose I apply myself to a complicated calculation and with much difficulty arrive at the result, I shall have gained nothing by my trouble if it has not enabled me to foresee the results of other analogous calculations, and to direct them with certainty, avoiding the blind groping with which I had to be contented the first time'.[1]

GENERALIZATION AND SIMPLICITY

When we generalize a result, we make it more useful. It may strike you as strange that generalization nearly always makes the

1. Poincaré, *The Future of Mathematics.*

result simpler too. The more powerful result is easier to learn than the less powerful one.

This may be illustrated by means of a very trivial puzzle, which runs as follows: One glass contains ten spoonfuls of water. Another glass contains ten spoonfuls of wine. A spoonful of water is taken from the first glass, put into the second glass, and the mixture is thoroughly stirred up. A spoonful of this mixture is then transferred to the first glass. Will the amount of wine in the first glass, at the end of this procedure, be more or less than the amount of water in the second glass?

The obvious way to go about this question involves the following calculation. After a spoonful of water has been put into the wine, the second glass contains 10 spoonfuls of wine and 1 of water, 11 spoonfuls in all. One spoonful of this mixture will therefore contain $\frac{10}{11}$ spoonful of wine, $\frac{1}{11}$ of a spoonful of water. After transferring it, the first glass will contain $9\frac{1}{11}$ spoonfuls water, $\frac{10}{11}$ spoonfuls wine. The second glass will contain $\frac{10}{11}$ spoonfuls of water, $9\frac{1}{11}$ spoonfuls of wine. The amount of wine in the first glass is therefore exactly the same as the amount of water in the second glass.

This being exactly the same might be an accident, but if you vary the conditions of the problem, you will find you always get equal amounts. If we begin with x teaspoonfuls of water and x teaspoonfuls of wine, the amount of wine that gets into the water still equals the amount of water that gets into the wine. Even if we make the glasses begin with unequal amounts, if we put x teaspoonfuls of water in one and y teaspoonfuls of wine in the other, and then follow the rest of the instructions as in the original problem, our calculation still shows that at the end the water in the first glass is equal in amount to the wine in the second glass.

Now this is a clear example of bad mathematical style. In a good proof, an illuminating proof, the result does not appear as a surprise in the last line; you can see it coming all the way.

This particular puzzle uses a kind of camouflage. *It tells you something which you do not need to know*, and by so doing distracts your attention from the real point. The unnecessary statement is 'the mixture is thoroughly stirred up'. The essential point is that we transfer one teaspoonful of liquid from the first glass to the second, and then we bring back one teaspoonful of liquid from the second to the first. It does not in the least matter what kind of liquid. All that matters is that, at the end, each glass contains the same amount of liquid that it did at the beginning. If this is so, the first glass must have received just enough wine to compensate for the water that it has lost; and of course the water it has lost will be found in the second glass. The amounts cannot fail to be equal; no fractions, no algebra need be employed.

The general statement of the puzzle would therefore be: we have a glass of water and a glass of wine, we carry out any series of operations with these liquids such that the total amount of liquid in each glass is at the end what it was at the beginning: then the amount of water that has got into the wine must equal the amount of wine that has got into the water.

But this is so obvious that it is hardly worth saying. It is much simpler than the calculations we had to do earlier. But the range of puzzles to which it applies is far greater; you could switch teaspoonfuls of liquid backwards and forwards between the two glasses as many times as you liked, and the principle would still apply.

The investigation of a problem therefore consists of scraping away all unwanted information, until only the essential facts remain. The less you are told, the easier it is to find a solution. A general theorem rarely says anything complicated; what it does is *to draw your attention to the important facts.*

In elementary mathematics we have a hotch-potch of details. In higher mathematics, we attempt to isolate the various elements involved, and to study each by itself. Higher mathematics can be much simpler than elementary mathematics.

Perhaps the most famous example of simplification by generalization is Hilbert's Finite Basis Theorem. In 1868 Gordan had proved, by laborious calculations, a certain theorem which I will not here attempt to state. It aimed at showing that certain collections of polynomials, which had arisen in connexion with a particular theory, had a certain property. In 1890 Hilbert proved this result, very simply and without calculations. The advance was due to his throwing away 90 per cent of the information used by Gordan. He proved the result to hold, not merely for those particular collections of polynomials, but for any collection of polynomials whatever!

We have gone from the ridiculous to the sublime. Nothing could be more trivial and localized than the wine and water problem, nothing more profound and far-reaching than Hilbert's theorem. The fact that both can be covered by the same maxim, 'Greater generality and greater simplicity go hand in hand', is perhaps an example of the power of mathematical statements to bring widely separated objects under the same roof.

One cannot judge the importance of any mathematical investigation by the particular objects it discusses. Topology is an example. Topology is sometimes referred to as 'the rubber geometry' – the geometry of figures drawn on an elastic sheet. And so in a way it is; it does treat of the properties of such figures. But its importance derives from the fact that on a rubber sheet there are no fixed lengths. There cannot be any result like Pythagoras' Theorem. We can only make remarks such as: This curve is continuous: that one is broken into two separate parts. Continuity is the basic property in topology, and topology has something to say about anything which is capable of *varying gradually*. As there are very few things incapable of such variation, topology has a very wide influence; it is increasingly becoming the concern both of pure mathematicians and of technicians; some most remarkable results can be proved by it. It has the utmost generality and the utmost simplicity.

UNIFICATION

All the tendencies we have so far discussed operate to enlarge the subject matter of mathematics. To explore, to discover patterns, to explain the significance of each pattern, to invent new patterns resembling those already known – each one of these activities increases the bulk of mathematics. From the practical viewpoint, it becomes extremely difficult to keep track of all the results that have been discovered; and a vast litter of unconnected theorems hardly constitutes a beautiful subject. Both as a business man and as an artist, the mathematician feels the urge to draw all these separate results together into one.

The history of mathematics therefore consists of alternate expansions and contractions. A problem occupies the attention of mathematicians; hundreds of papers are written, each clarifying one facet of the truth; the subject is growing. Then, helped perhaps by the information so painfully gathered together, some exceptional genius will say, 'All that we know can be seen as almost obvious if you look at it from this viewpoint, and bear this principle in mind'. It then ceases to be necessary to read the hundreds of separate contributions, except for the mathematical historian. The variegated results are welded together into a simple doctrine, the significant facts are separated from the chaff, the straight road to the desired conclusions is open to all. The bulk of what needs to be learnt has contracted. But this is not the end. The new methods having become common property, new problems are found which they are insufficient to solve, new gropings after solutions are made, new papers are published; expansion begins again.

If it were possible to weld together the whole of knowledge into two general laws, a mathematician would not be satisfied. He would not be happy until he had shown that these two laws were rooted in a single principle. Nor would he be happy then; indeed he would be miserable, for there would be nothing more for him to do. But there is not the least likelihood of this state of stagnation arising. It is a property of life, a property without which life would be unendurable, that the solution of one problem always creates another. There always is, there always will be something to learn, something to conquer.

The way in which this happens can be seen from a remarkable unification which took place round about 1800, when it was found that the great majority of functions previously studied were particular cases of one very general function, the hypergeometric function. The theory of the hypergeometric function was then, and still is, a powerful device for drawing scattered pieces of information together. New ways of viewing the hypergeometric function were discovered. It was shown, among other things, that the special properties of the hypergeometric function were connected with the fact that it had three singularities – exactly what singularities are need not be explained just now. The functions with simple properties, then, were those with three singularities or less; it was with such functions that previous mathematics had been concerned. But this immediately raised the question – unsolved, to a large extent, to this day – what sort of properties would a function with *four* singularities have?

And so it will always be. If it could be shown that all existing mathematics was concerned with things having properties *A*, *B*, and *C*, mathematicians would immediately ask, what happens if an object only has some, or none of these properties? And they would be off again.

CHAPTER THREE

Pattern in Elementary Mathematics

> Our stability is but balance, and wisdom lies
> In masterful administration of the unforeseen.
>
> R. Bridges, *The Testament of Beauty*

Let us consider pattern for a moment from the lowest possible point of view, that is, for a person who simply wants to pass an examination. The most important question for the examinee is the character of the examiner; what are the examiner's interests? what qualities is he trying to test in the examinee?

Some examiners seem mainly interested in the ability of the student to carry out routine operations. Does the student know his multiplication tables? Can he use logarithms? Can he use any one of a hundred other stock methods? Doubtless it is necessary to test the student's knowledge of routine processes; but ever since being a boy I have classified such tests as *dull examinations*. They are very popular with the poorer teachers, since the teacher knows he has merely to drill his class on exercises 1 to 50.

Other examiners wish to encourage the more enterprising teachers, the teachers who are trying to convey not merely facts but the *feel* of the subject; they wish to test the enterprise, the imagination, the initiative of the students. And so they seek out problems which call for these qualities.

Some teachers feel this to be unfair; they think that pupils cannot be prepared for an *unpredictable* examination. But this is not true. Is a battle predictable? Is it therefore impossible to train military leaders? The training of officers is based (or should be) partly on those general principles which are common to all battles, and partly in developing initiative; the officer is placed in a variety of unforeseen situations in which he has to improvise. Exactly the same type of approach is possible in the peaceful training of mathematicians. An examination and a battle have much in common.

An examination paper may contain both routine and problem

questions. How is the student to know which is which? They need not look very different.

Take for example the simultaneous equations below.

$$\left.\begin{aligned} 127x + 341y &= 274 \\ 218x + 73y &= 111 \end{aligned}\right\}$$

There is a routine method for solving such equations. If we multiply the first equation by 73, the second equation by 341, and subtract the first result from the second, we find that x must be 17,849/65,067. This answer does not simplify; there can therefore be no way of avoiding arithmetical calculations. An examiner who set such a question could merely be interested in seeing whether the students knew the routine method, and had the persistence and the accuracy to carry through the arithmetic. (I assume the exact answer, as a fraction, is wanted. If only an approximate answer is required, the labour can, of course, be reduced – for example, by using a slide rule.) The value of y is equally complicated.

For contrast, consider the simultaneous equations

$$\left.\begin{aligned} 6{,}751x + 3{,}249y &= 26{,}751 \\ 3{,}249x + 6{,}751y &= 23{,}249 \end{aligned}\right\}$$

The numbers here are larger, but the problem is a much easier one. The problem is not intended to test routine work. There is a very simple way into it if the examinee can find how.

What clues are there to suggest that this question calls for an imaginative attack? There are certain clues in the choice of the numbers, but the most striking and significant clue is the pattern

$$\begin{matrix} \circ & * \\ * & \circ \end{matrix}$$

which appears on the left-hand sides of the equations.

These left-hand sides have in fact the algebraic form

$$ax + by \ldots \text{(I)}$$
$$bx + ay \ldots \text{(II)}$$

In looking at these expressions I find myself reasoning in the following somewhat idiotic fashion. If we were to re-christen our unknowns x and y, so that x was called y and y was called x, the expression (I) would become (II), and the expression (II) would become (I). So the expressions (I) and (II) are just as good as each other. It would therefore be unfair to do anything to (I) that we do not do to (II). Whatever steps we take ought to treat (I) and (II) alike.

What steps are there of this kind? The obvious step is to add (I) and (II). This is a perfectly symmetrical operation.

Are there any other symmetrical operations that we could do? We could of course multiply (I) and (II) together, but that would not be helpful towards solving the equations. We want an operation of the form $m(\mathrm{I}) + n(\mathrm{II})$. At first sight, it seems that to be fair we must take $m = n$. Otherwise, whichever expression gets the bigger number, the other expression can complain of injustice. There is however a possible solution. The equation $p = q$ is certainly symmetrical as between p and q. But if we try to write this equation with all the non-zero terms on one side we obtain $p - q = 0$. This appears unsymmetrical; we have taken an arbitrary choice, in deciding to assign the + sign to p and the − sign to q. But if we had taken the other decision, we should have written $q - p = 0$, which again expresses $p = q$. Accordingly, although the *expression* $p - q$ is unsymmetrical, the *equation* $p - q = 0$, being a form of $p = q$, must be regarded as symmetrical. We need not feel we are sinning against symmetry if, given two expressions of equal status, we subtract one of them from the other.

Returning to our original problem, we now have two operations that seem to have the symmetry appropriate to the problem; to add and to subtract the equations given. Carrying out these operations, we find

$$10{,}000x + 10{,}000y = 50{,}000$$
$$2{,}502x - 2{,}502y = 2{,}502$$

that is, $x + y = 5$ and $x - y = 1$, from which $x = 3, y = 2$.

Children can find real pleasure in the experience of solving such a problem. Not everyone will follow the somewhat fanciful path I have just described; but anyone who solves this problem, other than by sheer calculation, will in some way be responding to the pattern of the equations. The conquest of the problem depends on sensitiveness to pattern; this combination of a military and an artistic aspect is perhaps not realized to such an extent in any activity outside mathematics.

In adult problems, nature is the examiner. And here again it is of the utmost importance to determine whether any particular problem involves a routine slog, or whether it has some special feature that will make possible a simpler solution. Many practical problems, of course, can only be solved by routine methods; they lack pattern. This is particularly so if there is an element of *randomness* in them. In surveying, for example, a host of geological and historical causes have interacted to determine the positions of towns and the mountains between them. One does not expect any elegant relations between the distances on the map; calculation is inevitable. On the other hand, in a fundamental scientific problem, the structure of the atom or of the universe, one does (rightly or wrongly) expect to find an underlying simplicity; the basic theories of physics usually possess mathematical elegance; their applications to complex situations, needless to say, may not.

RECONSTRUCTING AN EXAMINER

Occasionally palaeontologists dig up a small fossilized bone and proceed to reconstruct the shape of an extinct animal. A similar activity is possible in regard to examiners, the questions set taking the place of the fossil bone.

A good examination question is not just a shapeless affair; it should contain some interesting design or some surprising result. Such questions are by no means easy to make up. Accordingly, an examiner who is doing research work will usually be on the look-out for some result that he can use in an examination question. Often, in fairly advanced work, some small piece of algebraic manipulation occurs, which can be detached from its context and set as a problem.

For example, some years ago students brought me the following question, which had been set in an examination paper, and which they found hard to solve, or at any rate to solve in a satisfying manner.

'Prove that, if[1]

$$\frac{ac-b^2}{a-2b+c} = \frac{bd-c^2}{b-2c+d},$$

then the fractions just given are both equal to

1. It is assumed that b and c are unequal. The text does not discuss this point, as it is not relevant to the main theme. What suggested the question to the examiner?

$$\frac{ad-bc}{a-b-c+d}.$$

This question has a very definite form, and obviously to hammer it out by a lengthy and shapeless calculation, while verifying the result, would bring one no nearer to the heart of the question. What interested me most was the question, how did the examiner come to think of this question?

The pattern of the question includes the following aspects. $ac-b^2=0$ is the condition for the three quantities a, b, c to be in geometrical progression. The numerator of the first fraction contains $ac-b^2$. A similar expression occurs in the numerator of the second fraction. Down below we have expressions $a-2b+c$ and $b-2c+d$ which are associated with arithmetical progressions, $a-2b+c=0$ being the condition for a, b, c to be in A.P. Again there is a kind of rule by which the denominators could be derived from the numerators; in the third fraction, for example, we have a and d multiplied on top, added below, i.e. the numerator contains ad, the denominator $a+d$. The negative terms are similarly related; on top we have $-bc$, down below $-(b+c)$. This rule applies equally well to the first two fractions; in the first fraction, for instance, $-b^2$ is $-bb$, and we find $-(b+b)$, that is, $-2b$, below it.

To invent a problem so knit together is almost impossible. One does not invent such things; one stumbles upon them. I was certain that the examiner had been *finding the condition for something*, and these fractions had arisen in the course of the work.

The way to begin the problem was fairly obvious, to bring in a new symbol, k, for the value of the fractions. The problem then can be stated as follows.

If $$\frac{ac-b^2}{a-2b+c}=k\ldots\text{(I)}$$

and $$\frac{bd-c^2}{b-2c+d}=k\ldots\text{(II)}$$

prove $$\frac{ad-bc}{a-b-c+d}=k\ldots\text{(III)}$$

To bring in such a symbol k is routine procedure, when dealing with the equality of several fractions. (See, for example, Hall and Knight, *Higher Algebra*, Chapter 1.)

What to do next was not at all obvious to me. I tried various methods which, though they led to proofs, did not satisfy me. I continued to think about this question in odd moments, and about a week later, hit on the following approach. Equation (I) can be put in the form $ac-k(a+c) = b^2-2bk$. Both sides of this equation are now crying out for an extra term k^2 to complete their pattern. This will 'complete the square' on the right-hand side, giving $(b-k)^2$, and give $(a-k)(c-k)$ on the left. So $(a-k)(c-k) = (b-k)^2$. That is to say, *equation* (I) *expresses the fact that* $a-k$, $b-k$, $c-k$ *are in G.P.*

Now we have the whole thing. Equation (II) shows that $b-k$, $c-k$, $d-k$ are in G.P. So $a-k$, $b-k$, $c-k$, $d-k$ are in geometrical progression. But if we multiply together the first and fourth terms of a G.P. the result equals the product of the second and third terms. (Let the G.P. be A, AR, AR^2, AR^3. Then $A \times AR^3 = AR \times AR^2$.) So we have

$$(a-k)(d-k) = (b-k)(c-k).$$

If we multiply this out, cancel k^2 and solve the resulting linear equation for k, equation (III) results.

Our conclusion therefore is that the examiner's researches had led him on some occasion to pose the question, 'What is the condition that four numbers a, b, c, d must satisfy, if, by subtracting the same number from each of them, a geometrical progression can be obtained?'

The moral of this is not confined to examination questions. It is meant to support the thesis, *where there is pattern there is significance*. If in mathematical work of any kind we find a certain striking pattern recurs, it is always suggested that we should investigate *why* it occurs. It is bound to have some meaning, which we can grasp as an idea rather than as a collection of symbols. It is extremely unsatisfactory to discover a theorem, and only be able to prove it by shapeless calculations. It means that we do not understand what we have discovered.

To find the significance of an algebraic formula may take a long time; there are usually so many possible methods of attack, and no way of telling which is the true one. I find that, in dealing with such problems, my brain has a delayed action; at first I can make nothing of the question; a day, or a week, or a month later, an inspiration comes. If students are required to solve such a problem within three hours, that is, under examination conditions, a considerable element of luck is brought in. The usual way of overcoming this difficulty is to put in front of the problem a piece of bookwork, or some simpler problem that will start the mind working in the correct direction.

RECONSTRUCTING TWO AUTHORS

Even more difficult than collecting enough good questions to fill an examination paper is the task of collecting enough questions to fill a text-book. The best way to get interesting questions in algebra, I believe, would be to read a large number of research papers, written over the last couple of centuries, and take note of all the algebraic results that have been proved incidentally in the course of advanced work. The most interesting results in the elementary algebra books probably did originate in this way.

If one looks through any text-book, a certain number of examples are without pattern. They are the kind of thing anyone could easily make up; they belong to the same class of approach as the arithmetic question 'Find $27 + 46 + 39$' or the algebra exercise 'Multiply $5x^2 - 3x + 7$ by $4x + 11$'. A certain number of such routine exercises are of course necessary. Other questions are of the kind that could not just be made up on the spur of the moment, and it is always interesting to try to guess how the author arrived at his problem.

For example, in Hall and Knight's *Higher Algebra* (which of course is not higher algebra in the modern sense) the question occurs, If $a = zb + yc$, $b = xc + za$, $c = ya + xb$ prove

$$\frac{a^2}{1-x^2} = \frac{b^2}{1-y^2} = \frac{c^2}{1-z^2}.$$

This is not a difficult exercise to do, but we are not at the moment concerned with how it is solved; we are concerned with how it was composed.

Unless an enormous coincidence has occurred, it was arrived at in the following way. Mr Hall and Mr Knight were discussing, not algebra, but trigonometry. Now certain algebraic questions arise in connexion with trigonometry. In algebra, as is well known, you cannot find 3 unknowns from only 2 equations. If you have 3 unknowns and 3 equations, you can (as a rule) find the unknowns. If you have 3 unknowns and 4 equations, generally speaking, an impossibility arises. For example, you might have the equations $x = 1$, $y = 2$, $z = 3$, $x + y + z = 0$, which contradict each other. 4 equations for 3 unknowns can only be solved if the 4th equation is a mere repetition of something we could have discovered for ourselves from the other three, like, for example $x = 1$, $y = 2$, $z = 3$, $x + y + z = 6$. Then there is no trouble.

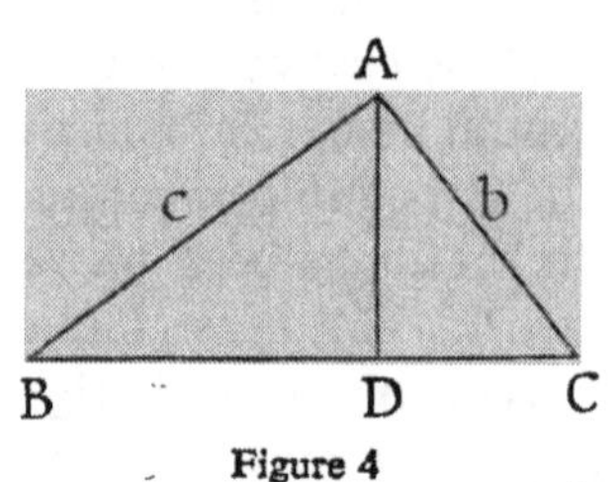

Figure 4

Now look at the question of solving a triangle (Figure 4), when the three sides a, b, c are given, and we are asked to find the angles A, B, C of the triangle. 3 unknowns, 3 equations will be enough; more than 3 may be an embarrassment. Trigonometry gives us eight.

First of all, if we drop a perpendicular AD on to BC, we have $BC = BD + DC$, and this shows

$$(1)\quad a = c \cos B + b \cos C$$

By dropping a perpendicular from B on to AC, or from C on to AB, we find two more equations, which I will call equations (2) and (3), but not write down here.

Three equations are quite enough to find three unknowns. From the point of view of algebra, we now have all we need. But trigonometry continues to pour its gifts upon us. The textbook calls our attention to the Cosine Formula

$$(4)\quad \cos A = \frac{b^2 + c^2 - a^2}{2bc}$$

and to the corresponding expressions for cos B and cos C, equations (5) and (6).

And finally there is the Sine Rule

$$\frac{a}{\sin A} = \frac{b}{\sin B} = \frac{c}{\sin C}$$

which contains two 'equal' signs, and gives us equations (7) and (8).

Now we know that we can find the angles of a triangle, and that these various results are not in conflict with each other. It must then be that equations (4) to (8) add no new knowledge, but are algebraic consequences of (1), (2) and (3).

This, as a matter of fact, is pretty obvious for equations (4), (5) and (6), which are simply what one gets on solving equations (1), (2) and (3) for cos A, cos B and cos C. But it is not so obvious that the Sine Rule follows. Here then is a result which is not too obvious, and will make an exercise in algebra. To get rid of the trigonometrical functions, we simply put $\cos A = x$, $\cos B = y$, $\cos C = z$. As $\sin^2 A + \cos^2 A = 1$, we shall have to put $\sin^2 A = 1 - x^2$. We square the Sine Rule equations above, and are thus able to translate the equations entirely into the language of algebra. We so arrive at the problem as stated in Hall and Knight.

In trigonometry, the cosine results are usually proved by one geometrical procedure, the Sine Rule by another. The fact that the Sine Rule is an algebraic consequence of the Cosine Formula is not stressed.

Even now, we have not given the actual algebraic reasoning for deducing one from the other. But we have seen that there *must be some way* of proving this result algebraically. If not, trigonometry contradicts itself.

TRANSLATION FROM ONE SUBJECT TO ANOTHER

The procedure of the previous section essentially is one of translation. We begin with a known trigonometrical fact, that the Sine Rule is not inconsistent with the Cosine Formula. We translate into the language of algebra, and find that it must be possible to deduce one set of equations from another. We have been led from a familiar fact (familiar at any rate to anyone who has done school trigonometry within the last two or three years) to an unfamiliar fact. We have not merely gained a new result; we have made somewhat sharper our view of the old result. 'Not inconsistent with' has been sharpened to 'can be deduced from'.

Translation is a valuable exercise, because, before you can express a fact in a new language, you must be clear what the fact is.[1] At various places in this book we shall try to state various things in such a way that they could be explained 'to an angel over the telephone'; to explain geometry, for instance, without drawing any diagrams or invoking any physical ideas. Geometry, in fact, can be explained purely in terms of number. You may think this strange, if you have been accustomed to think of geometry as the study of the shapes of objects. But mathematics deals with patterns; it is not concerned with the particular material in which the pattern is realized. Our ability to explain geometry to the angel means simply that we can, by means of numbers alone, exhibit the same *patterns* as those that occur in geometry. Like the wireless commentators, we could explain to the angel the progress of a football match, by giving the score and saying, 'Square 7'. We could not however explain to it what being kicked on the shin felt like.

As an example of translation, I will try to translate into algebra the fact that $\cos^2 \theta$ never exceeds 1.

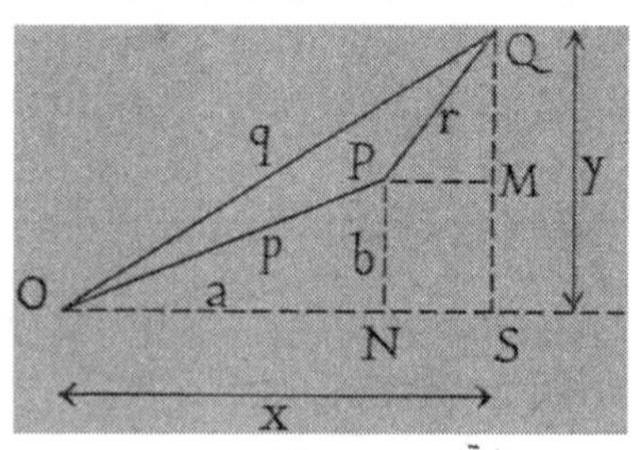

Figure 5

To get away from geometrical ideas we will make use of graph paper. In Figure 5 let O be the origin, P the point (a, b) and Q the point (x, y). This gives us a triangle OPQ, and anything there is to be said about this triangle can now be said in terms of the numbers a, b, x, y, that is to say, in terms of algebra. It will be convenient to use p for the length OP, q for the length OQ, and r for the length PQ. Of course p, q, r can be expressed in terms of a, b, x, y; by applying Pythagoras to the triangles ONP, OSQ and PMQ we have the very well known results

$$p^2 = a^2 + b^2$$
$$q^2 = x^2 + y^2$$
$$r^2 = (x-a)^2 + (y-b)^2$$

1. This is not a plea for compulsory Latin.

If the angle between OP and OQ is called θ, we can find $\cos\theta$ from the formula $2pq\cos\theta = p^2 + q^2 - r^2$. On substituting for p^2, q^2 and r^2 from the three equations above, all the square terms cancel out, and only $2ax + 2by$ remains. So we have

$$pq\cos\theta = ax + by.$$

p and q contain square roots, if we express them in terms of a, b, x, y; however, as we want to express the fact that $\cos^2\theta$ never exceeds 1, we shall be squaring anyway. We thus find that $(ax + by)^2$ cannot exceed p^2q^2, that is, $(ax + by)^2$ never exceeds $(a^2 + b^2)(x^2 + y^2)$.

This is a purely algebraic result; we may express it by saying that for real numbers, $(a^2 + b^2)(x^2 + y^2) - (ax + by)^2$ is never negative.

Why is it never negative? The above statement corresponds roughly to the trigonometrical statement that $1 - \cos^2\theta$ is never negative (for real angles θ). But this expression is the same as $\sin^2\theta$; we are familiar with the fact that squares of real numbers are never negative.

We are thus led to expect that the algebraic expression above may very well be a square. And so it is. It is no great labour to multiply it out completely. Of the seven terms so obtained, four cancel. Those remaining are equal to $(bx - ay)^2$. This can never be negative, and we have thus proved our algebraic result directly without any appeal to trigonometry.

Our result may be written

$$(a^2 + b^2)(x^2 + y^2) = (ax + by)^2 + (bx - ay)^2.$$

This identity holds for all numbers a, b, x, y.

In Chapter 4 we shall return to this identity, and see in what ways it can be generalized.

KNIGHT'S TOUR IN CHESS

While we are still thinking about translation, another example may be given, which illustrates one very important use of translation, namely, to put a problem into a form where the answer can be seen at a glance. Such translation does not change the essential pattern of a problem; from a very abstract mathematical viewpoint one might say that it did nothing at all: but for us as human beings it is most valuable, since it changes the unfamiliar into the familiar.

Drawing the graph of an algebraic function illustrates my meaning; a graph can be taken in by the eye at a glance. Here we are translating from algebra to geometry. Very often we translate in the opposite direction, from geometry to algebra; several examples will be found in this book. And we may also translate from geometry to geometry, from an unfamiliar type of problem to a familiar one.

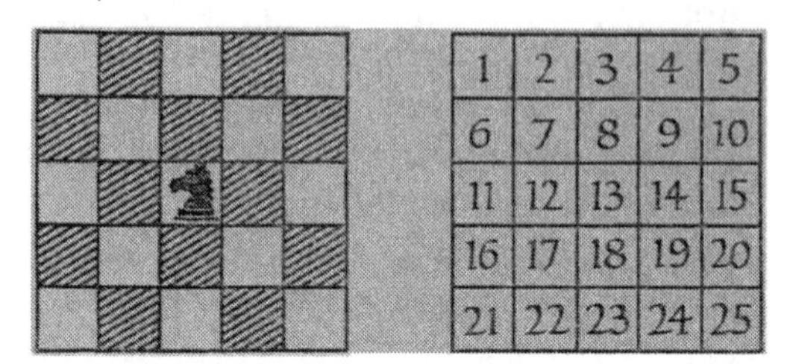

Figure 6

Consider the puzzle of the knight's tour. A knight is placed in the centre of a square board of 25 squares (Figure 6). The knight is to move in such a way that it visits each square once and once only.

I find it very difficult in thinking about this puzzle to see clearly in my mind whether or not the knight is getting into difficulties when he has made, say, a dozen moves: I cannot tell by looking at the squares still unvisited whether they will join together in a single chain of knight's moves. Can we somehow restate the problem so that we shall be able to see more easily what we are doing?

From the point of view of the knight, the squares 2 and 13 are neighbours. He can jump from 13 to 2 in a single move. But, from his point of view, 12 and 13 are not neighbours; he cannot pass from one to the other in a single move. So, if we are only concerned with the problem of the knight's tour, we can forget the actual shape of the chess board. If we want to bring out the things which are important for the task in hand, we must draw a diagram of the twenty-five squares in such a way that 2 is close to 13, while 12 is not so close. (In fact, the knight needs three moves to pass from 13 to 12.)

The resulting diagram is shown in Figure 7. If a knight can pass from one square to another in a single move, the corresponding numbers are joined by a straight line. The places where these lines cross have no significance. In a diagram drawn on paper it seems impossible to avoid such accidental crossings. They could be avoided in a three-dimensional model, with wires joining the points; these wires could be bent so as to avoid each other.

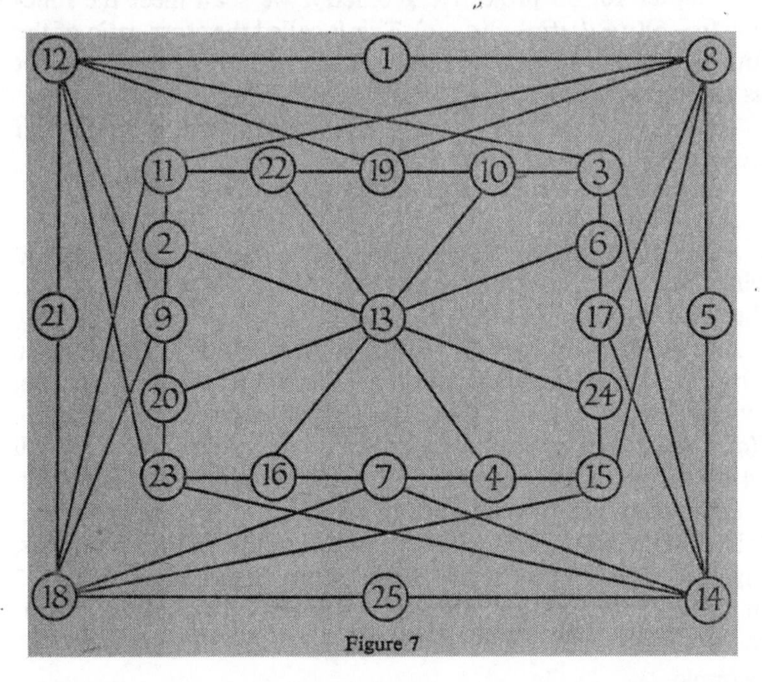

Figure 7

The drawing of such a diagram is not quite as simple as it looks. Some care is needed if one is to avoid a tangled network of lines. The main principle used was respect for symmetry. 13 had to be in the centre, and the others arranged round it so as to preserve the symmetry of the chess board.[1]

It is now very easy to choose a path for the knight. He can, for example, go from 13 to 10, then go right round the 'Inner Circle' (19, 22, 11, 2, 9, 20, 23, 16, 7, 4, 15, 24, 17, 6, 3), pass to the 'Outer Circle' and go round this (12, 21, 18, 25, 14, 5, 8, 1).

We need not go into the theory of knight's tours. The example

1. Mr W. H. Joint, Principal of Nuneaton Technical College, suggested the arrangement used in this diagram.

is chosen not for the importance of the subject, but rather to illustrate how a suitable way of visualizing the essential features of a problem can simplify the task of finding a solution.

A UNIQUE PATTERN

In Chapter 10, on projective geometry, we shall meet the function $(a-b)(c-d)/(b-c)(d-a)$. This is called the cross-ratio of the four numbers a, b, c, d. As it is rather a lengthy expression, let us refer to it as $f(a, b, c, d)$, which is somewhat shorter.

The order of the letters a, b, c, d is important. For instance, if we wrote $f(a, d, c, b)$ that would mean

$$(a-d)(c-b)/(d-c)(b-a),$$

which is not by any means the same as $f(a, b, c, d)$. In fact, as you can easily see, it is the reciprocal of $f(a, b, c, d)$. If $f(a, b, c, d) = x$, then $f(a, d, c, b) = 1/x$.

Altogether, there are 24 orders in which four letters can be written, so that there are 24 different expressions we can form, by taking a, b, c, d in different orders; for example $f(a, c, d, b)$, $f(d, c, b, a)$, $f(d, b, a, c)$. If you have the patience to write all 24 expressions out, you will find that they are not all different. 6 different values occur, each of them 4 times. You can check this algebraically; or, if you like, you can take particular values, say $a = 0$, $b = 1$, $c = 2$, $d = 3$. By taking the numbers 0, 1, 2, 3 in different orders, you can get the values $-\frac{1}{3}$, -3, $\frac{1}{4}$, 4, $\frac{3}{4}$, $\frac{4}{3}$ for the function; each value can be got in four different ways, for example

$$-\tfrac{1}{3} = f(0, 1, 2, 3) = f(1, 0, 3, 2) = f(2, 3, 0, 1) = f(3, 2, 1, 0).$$

The six values so found are all connected. If we take any one of them, and turn it upside down (take the reciprocal) we get another value in the set; if we take any value and subtract it from 1, we get another value in the set. For instance, if we start with $-\frac{1}{3}$ and turn it upside down we get -3. Subtracting -3 from 1 gives 4. Turn that upside down and get $\frac{1}{4}$. Subtract that from 1 to get $\frac{3}{4}$, and turn that upside down to get $\frac{4}{3}$.

Thus, if we are given any one value, we can find the others. If we start with x and turn it upside down we get $1/x$. Subtracting from one gives $1 - 1/x$, which may be written $(x-1)/x$. This turned upside down gives $x/(x-1)$. Subtract from 1 to get $1/(1-x)$ and finally turn this upside down to end with $1-x$.

You might think that by turning upside down and subtracting from 1 a few more times you might be able to get some new expressions. But you cannot. However many times you perform these operations you will still stay within the family circle of x, $1/x$, $1-x$, $1/(1-x)$, $(x-1)/x$ and $x/(x-1)$.

If then you were told that $f(a, b, c, d)$ had the value 5 say, you would know that $f(a, d, c, b)$ was $\frac{1}{5}$ and $f(a, c, b, d)$ was -4 (because $1-5 = -4$) *without needing to be told what a, b, c, d were.* For naturally the single equation $f(a, b, c, d) = 5$ is far from fixing the four quantities a, b, c, d.

The interest of this fact lies in its uniqueness. *No other function does what this one does.* Here, of course, we rule out trivial amendments to the function itself. For instance, the function $3f(a, b, c, d)-2$ has the same kind of property. But we regard this as the old function in disguise rather than as a new function.

Again, there are certain comparatively trivial ways of getting a similar property. We might consider a symmetric function, say $\varphi(a, b, c) = a+b+c$. The sum of three numbers does not depend on the order in which they are written. So, if we are told $\varphi(a, b, c) = 5$ we know that $\varphi(a, c, b) = 5$ too, and all the other ways of writing a, b, c would give 5 too. But this is too obvious to be interesting.

Slightly less obvious are functions such as

$$F(a, b, c) = (a-b)(a-c)(b-c).$$

If we exchange any two letters, this function changes sign. Thus

$$\begin{aligned} F(a, b, c) &= -F(a, c, b) = -F(c, b, a) \\ &= -F(b, a, c) = F(b, c, a) = F(c, a, b). \end{aligned}$$

It has only the two values x and $-x$.

However many letters are involved, we can always construct symmetric functions, like φ, which remain unchanged however the letters are shuffled, and functions like F, which only take two values when the letters are shuffled.

But if we want to have more than two values in the set, and to be able to predict all the values as soon as we are told one of them, we must work with *four* letters (such as a, b, c, d above), and our function must be the cross-ratio $f(a, b, c, d)$ or that function thinly disguised. In this sense, the number 4 and the cross-ratio $f(a, b, c, d)$ are singled out from all other numbers and functions.

An interesting thing is that the uniqueness of the function $f(a, b, c, d)$ can be proved without *making any calculations whatever*, that is to say, by a conversational type of argument. The proof is quite short, and depends on the theory of groups.

A unique function is nearly always worth studying. If certain towns can only be reached by passing a particular bridge, traffic on that bridge is likely to be heavy. In the same way, if certain properties are possessed only by one particular function, any problem which involves those properties must be solved by means of that function. The cross-ratio $f(a, b, c, d)$ does in fact occur in many branches of mathematics.

You may find it interesting – if you have not already done so – to verify, by elementary algebra, the property stated earlier in this chapter that if $f(a, b, c, d) = x$ then the 24 cross-ratios formed with a, b, c, d are the functions of x stated earlier. This involves a fair amount of work. Later, in Chapter 10, when we have learnt the geometrical significance of the cross-ratio, we shall be able to prove these results with very much less labour. You will appreciate the help that geometry gives to algebra very much more if you have worked at any rate a number of these verifications out for yourself by the more elementary method.

Incidentally, for anyone learning algebra, it is often more instructive to work one problem by three or four different methods than to work out three or four different problems. Working the same problem by different methods, one has a chance of learning how these methods compare with each other in shortness and efficiency. Thus an experienced outlook is built up.

CHAPTER FOUR

Generalization in Elementary Mathematics

> As mathematics passed the year 1800 and entered the recent period, there was a steady trend towards increasing abstractness and generality.... Abstractness and generality, directed to the creation of universal methods and inclusive theories, became the order of the day.
>
> E. T. Bell, *Development of Mathematics*

In Chapter 2 generalization was mentioned as one of the great processes leading to the growth of mathematics.

But generalization does not only make mathematics larger. It helps to tie the subject together. An unfamiliar result can be regarded as a generalization of a familiar one. This helps to give the new result a place in one's mind; it is tied on to the result already known.

For instance, at the beginning of school algebra, a child may be familiar with the fact that $x-a$ is a factor of x^2-a^2. Later it will meet the results that $x-a$ is a factor of x^3-a^3 and indeed of $x^n - a^n$ for every whole number n. The old, familiar result serves as a peg on to which to hang the new ones. It does not explain or prove the new results, but it helps the mind to accept them. Very often, one of the greatest difficulties of learning is not a logical difficulty at all. One sees every step, and admits that the proof is logical, but one is left with an obstinate feeling of not really knowing what the new result is, what it is all about.

For example, we meet sines and cosines at school. They are connected with circles, which we are thoroughly used to, and after working a few exercises we begin to feel that we know what sines and cosines are, even though we have to use tables to find the actual values of these functions. Later in life a mathematician meets Bessel functions. The properties of Bessel functions are learnt, the proofs are quite all right, but very often the student is left feeling that after all he does not know what a Bessel function *is*. He cannot see it in the same way that he sees a sine function.

Personally I was very much comforted when I discovered that a

Bessel function could be regarded as a generalization of a sine function; whether this idea will comfort other people in the same way I cannot predict. These are matters of psychology, not of logic. Anyway I give the idea for what it is worth.

In the theory of sound one meets the fact that when a stretched wire or string – a piano string, for instance – is vibrating, its shape will (in certain conditions) be a sine curve. Now of course a piano string is made of uniform wire, and has the same tension throughout. But it would be possible to have a string of variable thickness, and with different tensions in different parts of it. Variable

Figure 8

thickness is simply a question of manufacture; variable tension could be got, for example, by hanging the string vertically, or in various other ways which we need not now go into. Moreover, both things could be brought about gradually. We could begin with our ordinary piano string, and gradually coat it with extra material until it was thicker in some parts than in others. By means of rather complicated apparatus we could produce gradual variations of tension. The shape of the vibrating string would gradually change from a sine curve into some other shape. The Bessel function is one of the shapes that could be obtained in this way. Moreover, in this gradual process, a number of properties of the sine curve would be preserved, and we recognize certain properties of the Bessel function – for instance, its wavy graph – as coming from those of the sine.

As a matter of fact, the shape of a vibrating drum can be given by Bessel functions.

Indeed by means of a loaded piano string, one can obtain not only Bessel functions, but most of the functions that a mathematical physicist meets in his work. All of them can, in this respect, be regarded as generalizations of the sine. And it is a great help in learning about these functions to find very similar properties turning up again and again.

AN ALGEBRAIC GENERALIZATION

There is a section of algebra, taken in schools in some countries and in first year university work in others, which serves as an excellent illustration of generalization.

This section might be thought quite unsuitable for treatment in a popular work; (i) it is at or near university level, (ii) it involves, or can involve, very long algebraic expressions, (iii) students do not find it particularly easy.

Nevertheless, I intend to deal with it now. The above objections are met by the following considerations, (i) only an elementary knowledge of algebra is needed, and no long calculations occur, (ii) although the expressions are long, they have a very definite pattern, which the mind can grasp, (iii) the whole thing grows from a single idea, which is usually not mentioned in the text-books.

The single idea is the following; try to generalize the fact that there is one, and only one, straight line joining two points.

Now of course this is, as it stands, a geometrical result, and we might begin looking for a geometrical generalization. Someone may suggest the fact that a circle can be put through any three points; someone with a little wider knowledge may know that a parabola can be put through four points, and a conic section through five. But there we run into something of a dead end; nothing springs to mind as going through six points.

Accordingly we do what we did in Chapter 3, we try translating into algebra. That is easy enough. It is well known that any straight line (except a vertical one) can be got by choosing the constants m and c suitably in $y = mx + c$.[1] A point is specified by a pair of numbers. Our geometrical result when translated becomes 'it is always possible to choose m and c so that $y = mx + c$ passes through two given points (a, p) and (b, q).' We should add that a and b must not be equal, for then they would lie on the same vertical line, and the gradient m would have to be infinite.

Instead of saying that the line 'passes through' the points we could write the equations

1. Compare *M.D.*, Chapter 9, the section 'Mathematicians and Graphs', and the examples following that section.

$$p = ma + c\ldots\text{(I)}$$
$$q = mb + c\ldots\text{(II)}$$

These state that the expression $mx + c$ takes the value p when $x = a$, and the value q when $x = b$.

Already an algebraic aspect of the problem has appeared. We have two quantities, m and c, at our disposal, and there are two equations that they have to satisfy. But we are quite used to solving two equations for two unknowns.[1] We expect, given two equations for two unknowns, to be able to solve them. This suggests that if we want to put a curve through three points, we ought to find an equation with three unknowns, for example $y = gx^2 + hx + k$, where g, h, k are at our disposal. Now this idea is in fact correct, but we have not yet proved it.

Unfortunately, it is possible to have three equations in three unknowns with no solution. Consider for instance the equations

$$g - h = 1$$
$$h - k = 2$$
$$k - g = 3.$$

These have no solution. The first tells us that g is bigger than h, the second that h is bigger than k, the third that k is bigger than g; it is impossible to find three numbers for which this is true!

Actually, exceptional equations of this type do not arise from our particular problem; at any rate, they do not arise if we try to find $y = gx^2 + hx + k$ to pass through three points (a, p), (b, q), (c, r), where a, b, c are all different.[2] They would arise if, say, $a = b$. Suppose we tried for example, to find a formula of the type just considered to fit the table

x	1	1	3
y	2	4	5

This would mean that, when we substituted $x = 1$ in the expression $gx^2 + hx + c$ we should have to get the value 2 *and also* the value 4; and of course the same expression cannot have two different values for $x = 1$.

This argument incidentally shows that there must be exceptional systems of equations, which have no solutions; such

1. *M.D.*, Chapters 7 and 8, for instance. Chapter 8 is very close to our present topic.

2. This can be shown by means of determinants, which are discussed in Chapter 9.

equations must arise whenever we make ridiculous, self-contradictory demands. They are mathematics' way of saying, 'No'.

Fortunately for scientific workers, who often have to fit curves to points given by experimental results, the insoluble equations occur *only* when we ask for something obviously ridiculous. In this problem there are no unreasonable catches. If we take *any* three points, corresponding to different values of x, we can in fact put a quadratic graph through them. And it goes on in this way. To four points we can fit a cubic, to five a function of the fourth degree, and so on, indefinitely.

But this is still unproved, unless we are going to use the theory of determinants, to which we have not yet come.

Quite apart from proving that there *is* a function, it would be good to know what it is. If we can actually find the functions that solve the type of problem we are considering, this will automatically overcome the difficulty of proving that such functions exist; it will also save us solving equations every time we need such a function to fit experimental data.

THE SEARCH FOR PATTERN

How are we to look for functions that will do what we want? In such cases, it is usually wise *to take the simplest possible example and examine it carefully for hints of what happens in the more complicated cases.* This is a rule of general value; if you cannot solve some problem, make up for yourself the simplest problem of the same kind that you can devise, and see if it suggests anything.

In our problem, the simplest case is the one we started from, that of fitting $mx + c$ to two points. We can without difficulty solve the equations (I) and (II) given earlier, and find m and c in terms of the other letters. If this is done, the result is

$$m = \frac{p-q}{a-b}, \; c = \frac{aq-bp}{a-b},$$

and the line is

$$y = \left(\frac{p-q}{a-b}\right)x + \frac{aq-bp}{a-b} \ldots \text{(III)}$$

The algebra used in doing this is quite simple routine work. For anyone not in practice with algebra it may however represent quite an effort. Such a reader can follow one of two courses. The first is to take the result just given on trust, and go on with the main argument, which does not in any way depend on the details of the calculations used in finding this result. Afterwards, if desired, the calculations can be checked. This procedure leaves the brain fresh to deal with the main argument.

But some people can only get the 'feel' of the work by actually carrying out the calculations. For these, the wise plan is to go through the calculations first. If, when this has been done, the reader is fatigued, the best thing is to put the work aside, and return on a later occasion to the main argument, when the mind is again alert.

Having found the result (III), what can we learn from it? What hints does it give about the general result?

At first sight, not much; the algebraic expressions above look pretty untidy.

We notice perhaps that $a - b$ is the only expression that occurs down below; it is in the denominator both of m and c. And this is not surprising. As we saw, when $a = b$, the problem is an unreasonable one. Its solution must become meaningless when $a = b$, and it does just that. When $a = b$, everything becomes infinite. $y = \infty x + \infty$ is a pretty useless formula!

This perhaps is a hint that when we have three values of x, namely a, b, c, the only things occurring in the denominators will be $a-b$, $b-c$, and $c-a$.

For the rest, there is not much to suggest what the corresponding quadratic function through three points will look like. Very likely, if I were investigating this question for the first time, I would go on to work out the next case – the quadratic through three points – and see if I could draw any morals from that one. The algebra of course would be heavier, and the answer probably still more of a muddle, but one might detect some pattern in it.

As a matter of fact, there is a feature of the function we have already found. It is quite a striking feature, though I would not blame anyone for not seeing it.

A question to ask in studying an algebraic expression is, 'In what way does each symbol, taken separately, enter into it?' For example, the complicated expression $b^2c^3 + ac^5 - 7ab^4 + bc$ is simple in respect to the symbol a. If we regard everything but a as constant, for instance if we put $b = 1$, $c = 2$ it reduces to the linear expression $25a + 10$; and this would be so whatever constant values we took for b and c, we should simply get a linear expression in a.

Now the expression in equation (III) above is not particularly simple in the way in which a and b enter into it, but it is simple in regard to p and q. If we put in particular values for the other letters, say $a = 2$, $b = 1$, $x = 5$ we find it boils down to $4p-3q$. Whatever values we give to a, b, x we always find we get a linear expression in p and q, of the form $Ap + Bq$, where A and B are constants. This suggests that we rearrange equation (III), so that all the terms containing p are grouped together, and all those containing q are grouped together.

If this is done we arrive at the result

$$y = p\left(\frac{x-b}{a-b}\right) + q\left(\frac{a-x}{a-b}\right) \ldots \text{(IV)}$$

This is beginning to have some shape. To see how the shape works it is well to go back to our original problem. We were looking for a function, such that when $x = a$, $y = p$ and when $x = b$, $y = q$. How does the above expression manage to satisfy these requirements?

If we put $x = a$, the right-hand side above becomes $p \times 1 + q \times 0$. If we put $x = b$, it becomes $p \times 0 + q \times 1$.

This shows us how the formula works. Expression (IV) contains both p and q. But when $x = a$, y is to be simply p; q must be blotted out somehow. The formula achieves this by having next to q a bracket that becomes 0 when $x = a$, so that no q appears in the answer. On the other hand, when $x = b$, y should contain q, and contain it exactly once. So when $x = b$, the bracket takes the value 1.

The bracket that goes with p works the other way round. It has to be 0 and blot out p when $x = b$, and be 1 when $x = a$.

Suppose we call these brackets $f(x)$ and $\varphi(x)$ for short. Then $y = pf(x) + q\varphi(x)$.

In functional notation, what we have just said will read as follows; $f(a) = 1$, $f(b) = 0$, these two equations ensure that p is in the answer when $x = a$ and is not there when $x = b$; $\varphi(a) = 0$, $\varphi(b) = 1$, these two ensure that q is not there for $x = a$ but is there for $x = b$.

CONSTRUCTION OF THE DESIRED FUNCTIONS

Now we have an idea that can be applied to the more general problem. We will use it to find the quadratic such that when $x = a$, $y = p$; when $x = b$, $y = q$; when $x = c$, $y = r$.[1]

We begin by supposing $y = pu(x) + qv(x) + rw(x)$. $u(x)$, $v(x)$, $w(x)$ are three functions which will have the duty of seeing that p, q and r appear only when they are wanted. p is not wanted when $x = b$, nor when $x = c$. So $u(x)$ must be 0 when x takes the value b or c. That is easily arranged. If $u(x)$ contains the factor $x - b$, it will be zero when $x = b$. If it contains $x - c$ it will be zero for $x = c$. So we suppose it to contain both these factors. Now y is to be quadratic only, so we do not want $u(x)$, $v(x)$ and $w(x)$ to be anything above the second degree in x. Accordingly $u(x)$ can only contain the variable factors $x - b$ and $x - c$. There is however nothing to stop it having a constant factor as well. So we may suppose $u(x) = k(x-b)(x-c)$. This function will serve perfectly to shut out p when x is b or c, but it also has the job of admitting just one p when $x = a$. We want $u(a)$ to be 1. This fixes k. We must have $1 = k(a-b)(a-c)$, so

$$k = \frac{1}{(a-b)(a-c)}$$

You remember, we were expecting $a - b$ and $a - c$ downstairs.

Accordingly $$u(x) = \frac{(x-b)(x-c)}{(a-b)(a-c)}.$$

This has a recognizable pattern, and without any more working we can see that $v(x)$ should be taken as

$$\frac{(x-a)(x-c)}{(b-a)(b-c)}$$

and $w(x)$ as

$$\frac{(x-a)(x-b)}{(c-a)(c-b)}.$$

Putting all this together, we have our final result

$$y = p\frac{(x-b)(x-c)}{(a-b)(a-c)} + q\frac{(x-a)(x-c)}{(b-a)(b-c)} + r\frac{(x-a)(x-b)}{(c-a)(c-b)}$$

1. The c here of course has no connexion with the symbol c used earlier in $mx + c$.

If you like, by putting in the values a, b and c in turn for x, you can check that this does, in fact, give the desired values p, q, r.

Moreover, if you want to go on to the next problem, that of finding the cubic to fit four given points, no new ideas are needed. The pattern of the above expression is easily adapted. The formula for the cubic would be longer, but it would be no more difficult, essentially, than what we have just done.

The complexity of an algebraic expression, therefore, cannot be judged by the length of the expression, or the number of symbols it contains. It may be physically tiring, or mentally boring, to copy out a long formula; but if the longer formula brings in no new ideas, if its pattern has already been grasped by the mind, one should not regard it as being more difficult.

We could, if we had any sufficient reason for so doing, write the formula for the equation of the 17th degree through a given 18 points. It would be wearisome writing, but no new thought would be required.

GENERALIZATION AND RESEARCH

In reading mathematical publications, one is struck by the large number of papers which attempt to generalize a known result. Generalization is probably the easiest and most obvious way of enlarging mathematical knowledge.

One might think that the natural thing to do would be to think of some useful problem, and try to solve that. Indeed, much research begins with the attempt to solve problems. But a really difficult problem rarely yields to direct attack. One may cudgel one's brains for hours on end without getting any idea of how to begin to attack it. Doubtless it has a solution, but one cannot imagine what that solution will be like. Imagination needs something to feed on; one cannot produce a new idea out of a vacuum. And so one tends to reverse the process. One looks at a method that has worked well in the past; one tries to make that method more general; and finally one sees what problems the new results help to solve!

Often of course the desire to solve a problem guides one's choice of subject. One tries to generalize methods that have worked well on simpler problems of the same type.

An element of generalization is present in every new discovery.

One cannot help being influenced by the knowledge already in one's mind; the new discovery must have been suggested by something in that existing knowledge.

Sometimes, of course, in trying to generalize a familiar process one stumbles on something completely novel and unexpected. One may even discover that no generalization is possible; that the old process or result is of its own kind, unique.

These remarks may be illustrated by a result found in Chapter 3, that $(ax + by)^2$ never exceeds $(a^2 + b^2)(x^2 + y^2)$, all symbols standing for real numbers. This one result can lead to several different enquiries.

In one direction it can be generalized indefinitely. It is found that $(ax + by + cz)^2$ never exceeds $(a^2 + b^2 + c^2)(x^2 + y^2 + z^2)$, and one can find similar results with 4, 5, 6 . . . or any desired number of terms in the brackets.

Instead of trying merely to generalize the result one may seek to generalize the *method of proof*. A statement that one function never exceeds another is known as an *inequality*. The inequality given in Chapter 3 was proved by showing that the difference of the two functions was a perfect square. Can all inequalities be proved by a method of this kind? The answer is 'Yes'. If one expression[1] is less than another for all real values of the symbols, then the difference can be expressed as the sum of perfect squares.

In Chapter 3 we had the identity

$$(a^2 + b^2)(x^2 + y^2) = (ax + by)^2 + (bx - ay)^2$$

This identity shows a remarkable pattern, which leads us off in yet another direction. The first bracket, $a^2 + b^2$, is the sum of two squares. The second bracket, $x^2 + y^2$, is also the sum of two squares. And the whole of the right-hand side is the sum of two squares. This is often quoted in words – if the sum of two squares is multiplied by the sum of two squares the result is the sum of two squares.

Can this be generalized? Can we, in the statement above, replace the word 'two' throughout by any other number?

Well, of course, we can replace it by 'one', since

$$(a^2)(x^2) = (ax)^2.$$

1. By 'expression' is to be understood the simple type of function normally considered in elementary algebra.

This is not a very exciting result. A much more striking pattern comes on replacing 'two' by 'four'. The identity then is

$$\begin{aligned}(a^2 + b^2 + c^2 + d^2)(x^2 + y^2 + z^2 + t^2) = {} & (ax + by + cz + dt)^2 \\ & + (bx - ay + dz - ct)^2 \\ & + (cx - dy - az + bt)^2 \\ & + (dx + cy - bz - at)^2\end{aligned}$$

There is another identity in which eight squares multiplied by eight squares are equal to the sum of eight squares. I will not reproduce it here.

So we have had identities with 1, 2, 4 and 8 squares. What will the next one be? Obviously, one thinks, 16. But it is not so. The sequence breaks off. It has been proved that only for 1, 2, 4 and 8 can such an identity exist. Here then is something that cannot be generalized.

ALGEBRAS

Very closely related to the question just discussed is the topic of linear algebras.

Ordinary people are satisfied with plain numbers like 5 or $\frac{3}{4}$, but electricians and mathematicians have found it of great benefit to bring in a new symbol, i, together with the rule that i^2 may be replaced by -1. The resulting expressions of the form $x + iy$ can then be handled by exactly the same rules as the numbers of ordinary algebra.

It is natural to ask, 'It has been found advantageous to bring in this new number i. Why not bring in a few more letters, and see if good results follow?'

It sounds a promising line of enquiry, but the main result is negative. It has been proved that the extension of the real numbers to the complex numbers, $x + iy$, is the only extension that can be made if the rules of elementary algebra are to be preserved. Complex numbers are thus not merely useful; they possess a unique status.

If one wants to go beyond them, one must be willing to sacrifice something. The usual law to sacrifice is $ab = ba$. If we do not mind a times b becoming different from b times a, we can go on to quaternions, where we meet numbers such as (say) $3 + 4i + 5j + 6k$. The rules for combining the letters i, j, k are rather more elaborate than the rules for ordinary complex numbers. They are expressed by the following equations $i^2 = -1$, $j^2 = -1$, $k^2 = -1$; $ij = k$, $jk = i$, $ki = j$: $ji = -k$, $kj = -i$, $ik = -j$.

I give these equations in full in case you would like to use them to obtain the identity for four squares multiplied by four squares given in the previous section.

If you multiply out $(a + ib + jc + kd)(x - iy - jz - kt)$ by means of the rules for quaternions, you will find that you can get rid of i^2, ij, etc., and express the product in the form

$$(\ldots) + i(\ldots) + j(\ldots) + k(\ldots),$$

where the brackets contain certain algebraic expressions. The brackets are, in fact, exactly the same as the brackets occurring in the identity for four squares.

Quaternions themselves also mark a milestone. If you want to go beyond quaternions, you have to be prepared for still further sacrifices of ordinary algebraic laws. You must in fact be prepared to meet numbers p and q, neither of them zero, but yet having their product $pq = 0$, a thing which, of course, is impossible in ordinary arithmetic.

Ordinary numbers, complex numbers, and quaternions are thus marked off from all other algebras by a great gulf, and are in a class by themselves as objects of mathematical study. All three have, in fact, practical as well as theoretical importance.

CHAPTER FIVE

On Unification

> Throughout life he was always seeking for hidden connexions and an underlying unity in all things.
>
> Written of Friedrich Froebel
> (R. H. Quick, *Educational Reformers*)

In the last chapter we saw that every particular result can be thought of as a source from which generalizations spread out in all directions. The mind feels compelled to contemplate these generalizations, and at the same time feels the tremendous burden of this never ending variety. It becomes urgently necessary to compress knowledge again to manageable proportions, to unify this diversity of results.

One of the most satisfying moments in mathematical history is the instant when it appears that two departments of mathematics, until then regarded as separate and unconnected, are in fact disguised forms of one and the same thing.

One such striking piece of unification is within reach of the school syllabus. School mathematics seems to fall into two parts. On the one hand we have arithmetic, from which develops algebra, dealing with numbers. On the other we have geometry, and its development trigonometry, admittedly using numbers but mainly concerned with shapes. Trigonometry makes use of algebraic manipulation, but its foundation at any rate seems to lie in geometry and geometrical shapes, something quite distinct from the counting which is the foundation of arithmetic.

But then certain patterns begin to emerge in these two departments. In algebra we work out the powers of $(1 + x)$ and find

$$1 + x = 1 + x$$
$$(1 + x)^2 = 1 + 2x + x^2$$
$$(1 + x)^3 = 1 + 3x + 3x^2 + x^3$$
$$(1 + x)^4 = 1 + 4x + 6x^2 + 4x^3 + x^4$$

and thus arrive at the numbers of Pascal's triangle[1]

1. This name is given to the numbers listed in *M.D.*, Chapter 8, Table VI.

1 1
1 2 1
1 3 3 1
1 4 6 4 1

In trigonometry we meet the formula for tan $(A+B)$, namely

$$\tan(A+B) = \frac{\tan A + \tan B}{1 - \tan A \tan B}$$

From this formula, by putting $B = A$, tan $2A$ can be found. Then, by putting $B = 2A$, and using the last result, tan $3A$ can be found, and so on. If for shortness we write t for tan A the results are

$$\tan A = t$$

$$\tan 2A = \frac{2t}{1 - t^2}$$

$$\tan 3A = \frac{3t - t^3}{1 - 3t^2}$$

$$\tan 4A = \frac{4t - 4t^3}{1 - 6t^2 + t^4}$$

Here exactly the same set of numbers appear again. For instance, the numbers 1, 4, 6, 4, 1 that appeared in $(1 + x)^4$ appear in tan $4A$ in a wavering line

4 4
1 6 1

Admittedly there are minus signs before some of them. Nevertheless, the fact that the actual numbers of Pascal's triangle occur here is sufficient evidence of a common underlying pattern.

We might note this simply as an interesting oddity. But it was emphasized in Chapter 2 that pattern is significant; it is a symptom of some important relationship and calls for investigation. If we sought the basic reason for this common pattern we should eventually arrive at the famous equation[1]

$$e^{i\theta} = \cos\theta + i\sin\theta.$$

This relationship brings about the complete annexation of trigonometry by algebra. It becomes possible to *define* cos θ and sin θ as follows:

1. *M.D.*, Chapter 15.

$$\cos\theta = \frac{1}{2}(e^{i\theta} + e^{-i\theta}) \text{ and } \sin\theta = \frac{1}{2i}(e^{i\theta} - e^{-i\theta}),$$

purely algebraic definitions without any appeal to geometry. Trigonometry thus becomes purely a branch of algebra. All the properties of sines and cosines follow from the definitions given above, and can be proved more quickly and easily than by the usual elementary methods.

A skilful teacher can lead his pupils to discover for themselves both the Binomial Theorem and the relationship of trigonometry to algebra, thus reproducing in the classroom the discoveries of the seventeenth and eighteenth centuries, and giving the class the experience of witnessing and assisting a mathematical discovery.

The pleasure given by a unifying discovery is greatest when a person has struggled with the masses of undigested information in the old form, and is thoroughly familiar with these. Thus the unification of trigonometry and algebra above is appreciated by pupils who have worked through school trigonometry. One does not get the same satisfaction if one is presented with a lot of new information, which one only partly comprehends, and then hears of a unifying principle. For this reason many examples of mathematical unification which would be of interest to a mathematical student are not suitable for discussing in this chapter.

THE HYPERGEOMETRIC FUNCTION

At the end of Chapter 2 mention was made of a very remarkable example of unification, a single function which contained in itself very nearly every function that had previously been studied.

There are many different viewpoints from which the hypergeometric function can be regarded. Here we can discuss only one of these viewpoints – not the most instructive, but the one most capable of being shortly described – the hypergeometric function as represented by a series.

A very wide class of functions have the property of being described by means of series.[1] In elementary texts you will find such examples as the following

1. See *M.D.*, Chapter 14, especially the sections 'Other Series', 'The Dangers of Series', and 'The Series for e^x'.

$$\frac{1}{1-x} = 1 + x + x^2 + x^3 + \ldots$$

$$\frac{1}{(1-x)^2} = 1 + 2x + 3x^2 + 4x^3 + \ldots$$

$$\log_e (1+x) = x - \tfrac{1}{2}x^2 + \tfrac{1}{3}x^3 - \tfrac{1}{4}x^4 \ldots$$

$$\tan^{-1} x = x - \tfrac{1}{3}x^3 + \tfrac{1}{5}x^5 - \tfrac{1}{7}x^7 \ldots$$

$$e^x = 1 + \frac{x}{1} + \frac{x^2}{1.2} + \frac{x^3}{1.2.3} + \frac{x^4}{1.2.3.4} + \ldots$$

not to mention more formidable-looking results such as

$$\tfrac{1}{2}(\sin^{-1} x)^2 = \frac{x^2}{2} + \frac{2}{3}\,\frac{x^4}{4} + \frac{2.4}{3.5}\,\frac{x^6}{6} + \frac{2.4.6}{3.5.7}\,\frac{x^8}{8} + \ldots$$

This is only a very small sample from the series that have been discovered from time to time, with the following two properties (i) the function represented is one that occurs naturally in elementary mathematics, (ii) the series shows a definite pattern which allows us to write down further terms if we want them. For example, the next term in the series for $\tan^{-1} x$ would obviously be $+\tfrac{1}{9}x^9$; the next term in the last, rather complicated-looking series would be

$$\frac{2.4.6.8}{3.5.7.9}\,\frac{x^{10}}{10}.$$

I should perhaps mention to avoid any misunderstanding that the dots are intended to express multiplication. 2.4.6.8 means $2 \times 4 \times 6 \times 8$.

Anyone interested in seeing a great mass of examples of this kind will find them in the older calculus books, for example in Chapter 5 of Edwards' big *Differential Calculus.*

Particularly by studying the more complicated examples of such series, one can arrive at the conclusion that they are all related more or less directly to the following series

$$1 + \frac{a.b}{1.c}x + \frac{a(a+1)b(b+1)}{1.2c(c+1)}x^2 + \frac{a(a+1)(a+2)b(b+1)(b+2)}{1.2.3c(c+1)(c+2)}x^3 + \ldots$$

This series is denoted for short by the sign $F(a, b; c; x)$, because the series contains the three constants a, b, c, to which we can give values to suit ourselves, and the variable x. The pattern of the above series is not complicated – for instance, you would find no difficulty in writing down the terms of the series containing x^4 and x^5.

Some of the series we have given above can be obtained directly from this series. To take a very trivial example, if we make $a = b = c = 1$, we get $1 + x + x^2 + x^3 + \ldots$, the first series in our list. We can get the second series by putting $a = 2$, $b = 1$, $c = 1$.

The third series, for $\log(1 + x)$, we obviously cannot get directly by putting in values for a, b, c, because that series starts off with x, while $F(a, b; c; x)$ starts off with 1. However, if we put $a = 1$, $b = 1$, $c = 2$, we get something which is close to what we want. Each term is short of a factor x, and also we do not get the alternate + and – signs we need. But these defects are easily remedied, and you can easily verify the result

$$\log(1 + x) = xF(1, 1; 2; -x).$$

In the same way, we may meet series like $\tan^{-1} x$ in which only the odd powers of x occur. The appropriate result here is

$$\tan^{-1} x = xF(\tfrac{1}{2}, 1; 1\tfrac{1}{2}; -x^2).$$

To get e^x we have to allow some of the quantities to tend to infinity.

But none of these are very difficult operations; to bring in an extra factor x, to replace x by $-x$ or by $-x^2$, to let a constant become very large – all can easily be done. Regarding such operations as very simple, we see that the function $F(a, b; c; x)$ is remarkably adaptable. We can get almost any of the elementary functions we want from it without difficulty.

Besides the functions that occur in school work, there are many functions used by engineers or physicists – the Legendre polynomials and the Bessel functions, for example – which are particular cases of the hypergeometric function. In fact there must be many universities to-day where 95 per cent, if not 100 per cent, of the functions studied by physics, engineering, and even mathematics students, are covered by this single symbol $F(a, b; c; x)$.

What does this fact mean? That there are no other functions besides the hypergeometric type? Most certainly not; it is quite easy to write down functions of other types. The explanation lies in a different direction altogether.

RECOGNIZING OUR LIMITATIONS

Imagine farmers living in a country where no other tool was available except the wooden plough. Of necessity, the farms would have to be in those places where the earth was soft enough to be cultivated with a wooden implement. If the population grew sufficiently to occupy every suitable spot, the farms would become a map of the soft earth regions. If anyone ventured beyond this region, he would perish and leave no trace.

It is much the same with mathematical research. At any stage of history, mathematicians possess certain resources of knowledge, experience, and imagination. These resources are sufficient to resolve some problems but not others. If a mathematician attacks a problem which is completely beyond the range of the ideas available to him, he publishes no papers and leaves no trace in mathematical history. Other mathematicians, attacking problems within their powers, publish discoveries. Unconsciously, therefore, the map of mathematical knowledge comes to resemble the map of problems soluble by given tools.

But of course the discoveries themselves open the way for the invention of fresh tools. As the coming of the steel plough would change the map of the farmlands, so these new tools open up new regions of profitable research. But the new tools may take centuries to come, and while we wait for them, the frontier remains an impassable barrier.

Something of the sort seems to be the case with the hypergeometric function. It appears to be the limit of the kind of pattern we are able to recognize at present. If one goes just beyond its boundaries, everything seems formless. Beyond doubt, pattern exists there, but it is the kind of pattern we have not yet learnt to see.

I do not wish to imply that the hypergeometric function is the only function about which mathematicians know anything. That is far from being true. There are other fertile valleys with which the wooden ploughs of the twentieth century can cope; but the valley inhabited by schoolboys, by engineers, by physicists, and by students of elementary mathematics, is the valley of the Hypergeometric Function, and its boundaries are (but for one or two small clefts explored by pioneers) virgin rock.

CHAPTER SIX

Geometries other than Euclid's

I could be bounded in a nutshell and count myself a king of infinite space.

Hamlet

LEARNING TO FORGET

The main difficulty in many modern developments of mathematics is not to learn new ideas but to forget old ones. To take, for example, Einstein's theory of relativity; I do not believe that an angel, a disembodied creature without any experience of space or time, would find Einstein's various theories any harder to understand than the older views of the universe. The difficulty in grasping a new theory is that one tends to carry over to it habits of thinking which belong to the old theory. In the last 150 years, mathematical ideas have been in a continual state of flux. Traditional mathematical ideas have been closely examined and, again and again, have been found either meaningless or wrong. If there is any value in mathematical education as influencing one's general attitude to life, it probably lies in this training. The present age is a striking example of the chaos produced by the slow movement of men's ideas. To future ages – if there are any future ages, if we do not smash the whole basis of life in our blunderings – we shall most certainly appear savages and lunatics. But it is very hard for us to look at the present world and see it exactly as it is, or could be; we are bemused by the mists of the past.

Non-Euclidean geometry is one of the new ideas of mathematics; in fact, the first and most striking break with tradition. Before 1800, and I suppose still to-day for some schoolboys and schoolgirls, Euclid is the one true geometry, something certain and proved. As children we tend to think of 'Geometry' rather than 'a geometry'.

PHYSICAL AND MATHEMATICAL TRUTHS

It is necessary to make a distinction between something which is physically true and something which is mathematically true. Physics rests upon experiment. Physical truths are – for example – that the earth is (very roughly) a sphere of 4,000 miles radius, at a distance of about 90,000,000 miles from the sun. But one could very easily imagine these things being otherwise. One could imagine the earth being smaller or larger; its shape could be changed to a cube or a tetrahedron; there is no special reason why it should be just 90 million rather than 47 million or 132 million miles from the sun.

Mathematics is concerned with things that could not be otherwise without logical contradiction. It is concerned with how one thing follows from another. A jury may on occasion return a verdict that is logically correct but actually untrue. Suppose for example that a person in fact has committed a murder, but that insufficient evidence is available to prove guilt. The jury act correctly in returning a verdict of 'Not guilty'; the guilt has not been *proved*; whether the guilt exists is another question. The jury in fact act as mathematicians. If new evidence is brought to light after the trial, the jury are not to blame. In much the same way, one might say that a general acted reasonably in view of the information available to him at the time of the battle, even though the actual consequences were disastrous in view of certain facts which he did not and could not know.

The mathematical question then is not, 'Do the angles of a triangle in fact add up to 180°?' but 'Is it logically necessary that the angles of a triangle *must* add up to 180°?' If Euclid's were the one logically possible geometry, it would mean that if you were going to create a universe, that universe would have to have Euclid's geometry. Can we imagine a universe with a geometry different from ours?

There is of course also the question for the physicist: is the geometry of this universe in fact Euclid's?

Mathematically it is certain that other geometries are possible; physically it may well be that Euclid's geometry is not exactly true of this universe.

NUMBER OF DIMENSIONS

We may begin loosening up our ideas by thinking about the number of dimensions. We live in a space of three dimensions; you can go x miles to the East, y miles to the North, and z miles up. Your distance from where you started would then be s miles, given by the formula $s^2 = x^2 + y^2 + z^2$.

If you are now deprived of your balloon or helicopter, so that you can no longer rise from the surface of the earth, you are now confined to a space of two dimensions. You can go x miles to the East, and y miles North, and your distance is given by $s^2 = x^2 + y^2$.

If we limit you still further, by requiring you to ride in a truck on a straight railway, running East and West, you now move in one dimension only. You can go x miles to the East (if you travel West x will be a minus number), and the distance you have gone is given by $s^2 = x^2$, which I leave in this form to keep the analogy with the earlier two formulae. Finally, if you were chained to a post, you could not move at all: you are now in 0 dimensions.

To us it seems quite reasonable that there should be just these four possible spaces – points, lines, planes and solids. But let us see how it looks to an inhabitant of another universe. Suppose you have to explain the ideas of geometry to an angel over the telephone. By an angel I mean some creature with no physical experience at all. Length, colour, and so forth mean nothing to it. We must imagine it able to hear and speak, so that we can communicate with it, and we will suppose it to be extremely intelligent. But of course we cannot show it diagrams or sketches; for one thing, we are speaking to it over the telephone, and for another, it cannot see or touch. We are able to convey to it what number means. We go tap, tap, tap on the telephone mouthpiece and say, 'Those were three taps'. Being so intelligent, it soon has the idea of the whole numbers 1, 2, 3, 4, 5 ... and from these we explain fractions $\frac{3}{5}$ and $\frac{2}{7}$ and so forth, and we also manage to make clear the meaning of numbers like $\sqrt{2}$ and π and e. All of these of course have to be explained purely arithmetically: we cannot explain π, for instance, as the circumference of a circle divided by the diameter, since the angel does not know what a circle is, but we can give an infinite series for π.

Now we come to geometry. Obviously the only way we can explain it is by means of co-ordinate geometry, in which every geometrical fact is turned into a fact of arithmetic or algebra. So we start off. The angel does not know what *East*, *North* and *up* mean, so we do not bring these terms in. We start off, 'A line is made up of points. You cannot understand what we mean by points or lines, but anyhow, on a line, a point is fixed by giving a single number, x. In a plane, a point is fixed by giving a pair of numbers (x, y). In space of three dimensions, a point is fixed by giving three numbers (x, y, z). Then there is a thing called distance. This is measured by a number, s.

On a line $s^2 = x^2$.

In a plane $s^2 = x^2 + y^2$.

In three dimensions, $s^2 = x^2 + y^2 + z^2$.'

Here we stop. The angel is disappointed. It expected us to go on and say that in four dimensions a point was fixed by four numbers (x, y, z, t) and distance was given by $s^2 = x^2 + y^2 + z^2 + t^2$; and after this it expected to hear something about five, and six, and seven dimensions, and so on, indefinitely.

Mathematically, there is no obvious reason for stopping at 3 rather than any other number. In *this* universe, *North* is perpendicular to *East*, and *up* is perpendicular to both, but we cannot find a fourth direction perpendicular to all three. There is however no reason why a universe should not exist with four or five or six dimensions. We have got used to three, but that is not a reason.

Indeed, there is no need to go as far as 3. One could create a perfectly comfortable universe with only two dimensions; the people would be shapes moving about in a plane. They would have East and North, but no 'up'. You may ask, 'But what is above or below the plane?' There is no answer; this is a question you are not allowed to ask. It has no physical meaning in a plane universe. A dweller in four dimensions might just as well ask us, 'You can go East and North and up, but what happens if you want to go in the fourth direction, at right angles to all three?' We can only answer, 'There is no direction at right angles to all three.'

Later in this chapter we shall consider people living in the surface of a sphere. You may ask 'What is inside the sphere, what is outside it?' You must not ask this! The surface of the sphere is their universe. For them, nothing else has physical reality.

In fact, the whole nature of a universe would be changed if creatures in it gained the power to enter an extra dimension – that is, if the idea of a further dimension became a physical reality.

In a plane (Figure 9), for instance, a square constitutes a prison. If a square is drawn on the top of a table, and a coin placed inside the square, the coin cannot leave the square without passing through the walls of the square – that is to say, if the coin simply slides on the table's surface, in two dimensions. But if the coin is allowed to move into the third dimension 'up', it can go from the inside of the square to the outside of the square, *without* passing through the walls of the square.

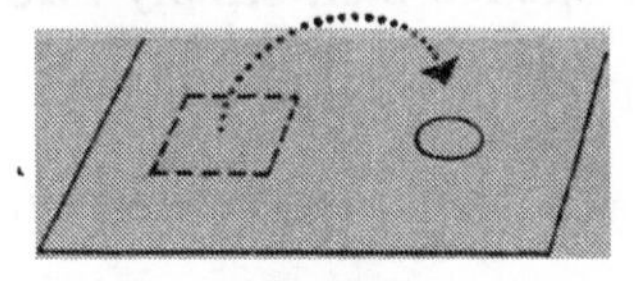

Figure 9

In the same way, if we in three dimensions acquired the power to travel in a fourth dimension, we could escape from a closed prison cell. Suppose for example you had the power to travel in time. If you were locked up in a fortress, you could travel back in time to before the fortress was built, walk half a mile, and then return to the present day. To the rest of us ordinary dwellers in three dimensions it would seem that you had disappeared from inside the fortress, and reappeared half a mile away.

But this is pure fantasy. We can imagine it, but we cannot do it. It has mathematical but not physical reality. In the same way, when we come to consider our universe that is the surface of a sphere, you must remember that only the surface is physically real to these people. To us as creators it may have meaning to talk of being inside or outside the sphere; philosophers in our spherical universe may speculate about such ideas – as I above speculated about journeys in time – but these are idle speculations; none of our creatures can, in actual fact, leave the surface of the sphere. That is their world, that is their reality.

It may be well to make clear that, as words are used among mathematicians at present, the mere fact of a universe having 4 or 5 or 6 dimensions would not make it non-Euclidean. True, it would be a different universe from the one Euclid thought about;

but the term non-Euclidean is reserved for another use. The central feature of Euclid's geometry is Pythagoras' Theorem. In the plane, Pythagoras' Theorem is simply the formula $s^2 = x^2 + y^2$ that we had earlier. In three dimensions, Pythagoras' Theorem takes the form $s^2 = x^2 + y^2 + z^2$. In four dimensions, it would be $s^2 = x^2 + y^2 + z^2 + t^2$; in five dimensions, $s^2 = x^2 + y^2 + z^2 + t^2 + u^2$; and so on. We should describe these geometries as 'Euclidean space of 4 dimensions' and 'Euclidean space of 5 dimensions', because they are built around Pythagoras' Theorem in an amended form. By a non-Euclidean geometry, on the other hand, we mean one in which Pythagoras' Theorem is no longer true. This is a much more profound change than simply throwing in another dimension or two.

MATERIALS FOR A NEW UNIVERSE

It may very reasonably be said that mathematicians are wasting their time thinking what the universe *might* have been like, instead of being good physicists and discussing what it *is*. The moon might have been made of green cheese, but the fact is of no significance. Equally, it could be argued, it is pointless to say that the universe might have had four or only two dimensions of space, when in fact it obviously has three.

In a story of Graham Greene's a lawyer questions a witness; 'You are sure this is the man you saw?' 'Yes.' 'Could you swear it was not his twin brother who is sitting at the back of the court?' After the witness had seen the twin, the witness admitted it was impossible to tell one from the other.

Mathematicians have done something the same with the universe. We were all willing to swear that we were living in Universe I, Euclid's universe. Mathematicians have produced two other universes, Universe II and Universe III, which are not identical with Universe I – indeed, they differ in very important respects – but the resemblance is sufficient to make us uncertain whether we live in I, II, or III. We are in the position of the child who ran out in a street of mass-produced houses, and could not remember which was its home.

Let us first consider the resemblances between Universes I, II, and III. In all of them there are rigid bodies, which are able to move about. By a rigid body I mean something like a brick or a

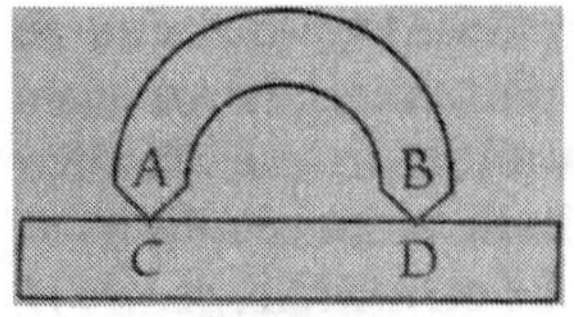

Figure 10

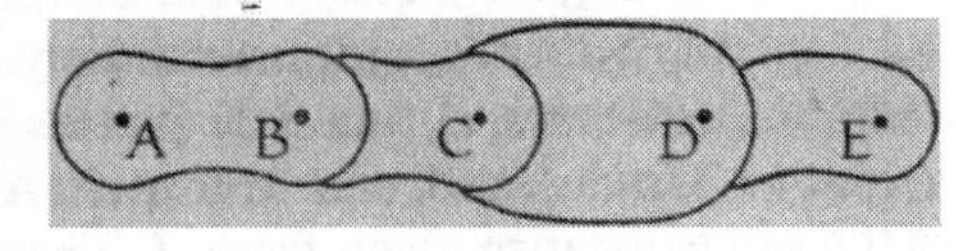

Figure 11

steel bar, not something like clay or putty. You would not make a ruler out of putty, or elastic.

To test whether bodies are rigid is fairly simple. In Figure 10, the points A and B of the horse-shoe shaped object just touch the points C and D marked on the ruler. I separate the objects and wave them about, and then bring them together again; if each time I do this, I find that the points A and B can be made to touch C and D, I conclude that the objects are rigid. I can then bring in the idea of *length*, and say that AB has the same length as CD.

We also assume, in each of the Universes I, II, and III, that bodies behave in the same way at all places. If AB has the same length as CD when I compare the objects here, AB and CD will have the same length if I take them both to America and measure one against the other there.

Two assumptions so far: (i) *rigid bodies, freely movable, making possible the definition of length,* (ii) *the properties of space the same at all points.*

By a line we shall understand the shortest distance between two points. We can get a line, physically, by making a chain of rigid bodies, as shown. If we pull A and E until they are as far apart as possible, the points B, C and D, at which the successive links of the chain are joined to each other, will lie on the straight line joining A to E (Figure 11).

We make assumption (iii), *there is only one line joining two points.*

You may notice that our rigid bodies allow us to measure angle as well as length. Figure 12 shows how to make a protractor. We have a rigid triangle, ABC. We swing it round to the position AB_1C_1, where AB_1 is in the same direction that AC was at first. Then we swing it round to the position AB_2C_2, and so on. If the

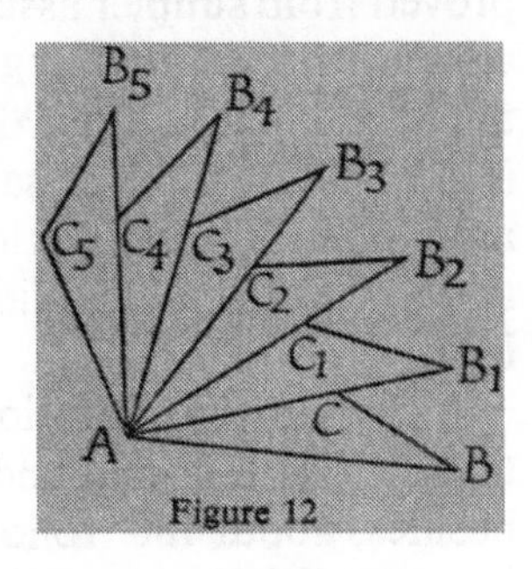

Figure 12

triangle ABC were chosen with a sufficiently sharp point at A, the lines AB, AB_1, AB_2, etc., would be very close together, and we could measure angles quite accurately. We can talk of right angles in Universes II and III, just as well as in our familiar Universe I. Circles, too, can be drawn. All the points $B, B_1, B_2 \ldots$ are the same distance away from A. They lie on a circle, centre A. All of this is quite homely and usual.

WHERE PEOPLE FIRST DOUBTED EUCLID

Most of the things Euclid assumes, or takes for granted without mentioning, are very 'reasonable' – regarded, that is, as a physical description of this universe. The ordinary man is prepared to agree to them. But one of the things Euclid assumes is not so

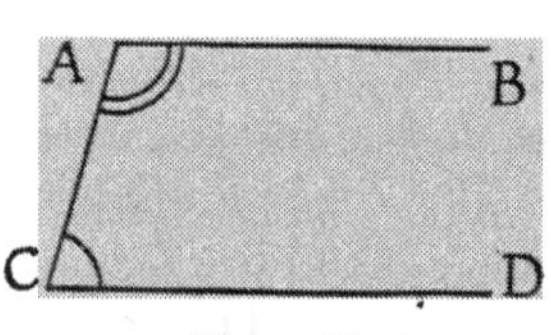

Figure 13

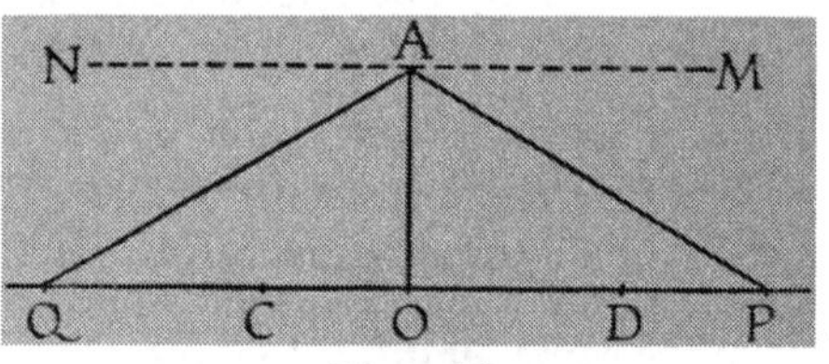

Figure 14

obvious. He assumes this: through a point A (Figure 13) one and only one line AB can be drawn parallel to CD, and the angles BAC and ACD then add up to 180°. If their sum is less than 180°, AB will meet CD somewhere to the right of C. If their sum is more than 180°, AB produced will meet CD somewhere to the left of C.

Now most of us would accept this as a theorem, something proved from simpler assumptions. But Euclid does not give it as a consequence of anything simpler. You must agree to this right at the beginning; without this, we cannot start. Euclid does not say it is obvious; he does say he cannot reduce it to any simpler or more plausible assumption. Over the centuries many people tried to do what Euclid had failed to do, to reduce this to some simpler idea.

A certain simplification was achieved. It is not hard to show that, if there is one and only one parallel line through A, the business about 180° follows from it. But can we assume that there is a parallel line, and only one? And how can we decide? To establish that two lines are parallel, you have to produce both of them to infinity, and see that they still do not meet; and infinity is a place where none of us have been.

Let us look a little more closely at what is involved. In Figure 14 *AO* is drawn perpendicular to the line *CD*. *P* and *Q* are two points on *CD*. *Q* is just as far to the left of *O* as *P* is to the right of *O*. *AM* is a line drawn to the right from *A*, so that the angle *OAM* is 90°. *AN* is drawn to the left, so that angle *OAN* is 90°. As these two right-angles add up to 180°, *NAM* is a straight line.

According to Euclid, if *P* moves off to the right along *CD*, the line *AP* will swing round closer and closer to *AM*; it will get as near to *AM* as you like, but will never quite reach it. On the left-hand half of the figure (which is symmetrical about *OA*), the line *AQ* similarly will get indefinitely close to *AN*, but will never reach it.

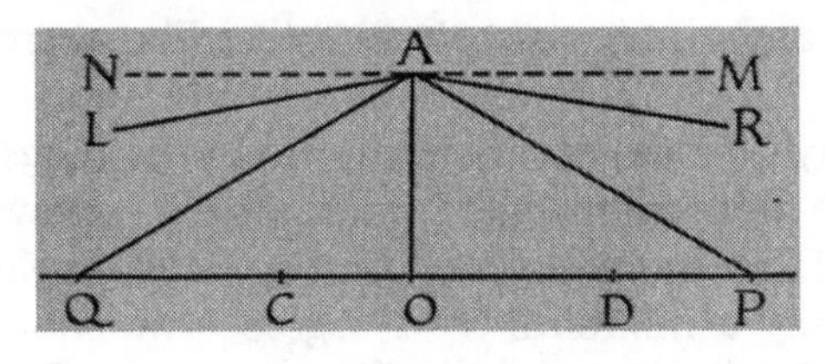

Figure 15

Euclid may be right about this. Let us call the possibility that he is correct *Possibility I.* It corresponds to Universe I.

What can happen if he is wrong? He says that *AP* approaches as close as you like to *AM* but never reaches it for any finite distance *OP*. This could be wrong in two different ways:

Possibility II. For a finite distance *OP* the line *AP* may actually reach the position *AM.*

Possibility III. On the other hand, it might be that *AP* could not approach indefinitely close to *AM* (Figure 15). That is to say, as *P* moves to the right, *AP* approaches, without ever actually reaching, the direction *AR*, which lies below *AM*. In the same way, as *Q* moved to the left, *AQ* would approach, without ever reaching, the direction *AL*, which lies below *AN*.

If you draw a figure to illustrate Possibility II, or look at the figure I have drawn for Possibility III, you will say that these figures look wrong. I agree; to people who have been brought up as we have, they do look wrong. For centuries mathematicians believed that they were impossibilities – although no logical inconsistency could be proved against either of them. About 1830, two mathematicians working quite independently – Lobachewsky in Russia and Bolyai in Hungary – published papers which admitted Possibility III as a reasonable viewpoint. In 1854, Riemann recognized Possibility II.

It is not surprising that Possibility III was seen before Possibility II, for Possibility II has a strange consequence. Under (II), the point P can move a certain finite distance, say k, to the right of O, and then OP coincides with AM. That is to say, NM meets CD at a distance k to the right of O. But by considering Q, by just the same argument we see that NM meets CD at a distance k to the left of O. Now two lines can only meet in one point. So the point you get by going a distance k to the right of O must be the same as the point you get by going a distance k to the left of O. In fact, the straight line must behave rather like a circle. But why not?

If Possibility II is taken, there are no such things as parallel lines. Any line drawn through A will meet CD somewhere.

Possibility III is not nearly so drastic. If the angle RAM was very small – say a millionth of a degree – we should have the utmost difficulty in distinguishing the geometry from Euclid's. But there would be an infinity of lines through A parallel to CD – any line in fact that made an angle of less than a millionth of a degree with AM. But, without very accurate measuring apparatus it would be very difficult to distinguish this little bundle of parallel lines from the single parallel line of Euclid.

A SPHERICAL GEOMETRY

I now want to consider the life of creatures whose whole universe is the surface of a sphere. This geometry is not a perfect illustration of Riemann's geometry – Universe II – but it throws a good deal of light on Universe II, it illustrates several logical points, and, since a sphere is so well known, it is very easy to see what is happening. It is useful to have a globe of the world, or a ball, at hand when reading this section, and also a small piece of plasticine or similar material.

First of all, the surface of the sphere does satisfy assumptions (i) and (ii). A rigid body can move freely on the sphere. If you mould your plasticine so that it just covers India on the globe, you will find the plasticine can then slip freely on the globe's surface. You can slide your plasticine outline of India until it lies over Europe, you can rotate it freely, and all without leaving the surface of the sphere. Assumption (ii) was that the properties of space were the same at all points – this is true of a sphere: any point of it is exactly like any other point.

If your plasticine sets hard, you have a rigid body. By moving it about the sphere, you can compare lengths at different places. It may interest you to go back to the earlier diagrams – for making a protractor, for instance, or of the ruler and the horseshoe – and see that all this argument still works on the surface of a sphere. We can speak of lengths and angles on the globe's surface. Lengths can be measured by a thread stretched on the globe. The meridian of Greenwich meets the equator at right angles.

What is a 'straight line' in this universe? It is the shortest distance between two points. As was emphasized earlier, the universe is limited to the surface of the sphere; the shortest distance means the shortest route on that surface – no tunnelling allowed! To us, looking at the globe from outside, the straight path appears curved; the equator, for example, gives the shortest route between two points on itself; the shortest way from London to the Gold Coast is to go straight down the meridian of Greenwich. For the purpose of the earthbound, these are 'straight lines'; they are given by the definition, they are the shortest paths.

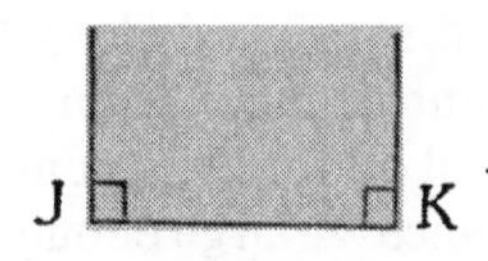

Figure 16

A defect of this model, from the point of view of geometry (II), is that 'straight lines' meet in two points; for example, any two meridians meet both at the North Pole and the South Pole. But so long as we keep to a limited part of the world this complication will not arise.

The globe illustrates well one aspect of geometry (II) – there are no parallel lines. Any two 'straight lines' meet. In Euclid's geometry, if you take a straight line *JK* and make right-angles at both ends you get a pair of parallel lines (Figure 16).

But try this on the surface of the globe. Take *J* and *K*, for example, on the equator (Figure 17). The two 'lines' at right angles to *JK* will be meridians, and these will meet at the North Pole.

An interesting objection to the statement that there cannot be parallel straight lines on the surface of the globe is 'But what about railway lines?' The whole point of rails is that they should remain a constant distance apart, to leave room for the axle. How then can they meet?

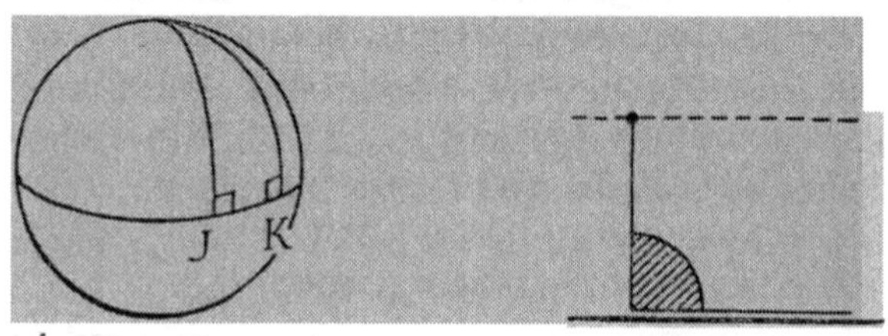

Figure 17 Figure 18

Let us put this in rather more geometrical terms. If we had one rail of a railway track laid, we could scrape out the track where the other rail ought to go by having a rod so arranged that it could slip along the first rail, but always remaining perpendicular to that rail (Figure 18). In fact (in Euclid's geometry) 'the locus of a point, which moves so that its perpendicular distance from a given straight line is constant, is a parallel straight line'. What happens to this construction on the globe?

Let us imagine the first rail laid along the equator. The second rail is to be a constant distance from the equator. Any of the parallels of latitude will do for the second rail. We might take the Arctic Circle for instance. A large enough vehicle could run with one pair of wheels on the equator, the other on the Arctic Circle. And certainly the equator and the Arctic Circle never meet. Here then are parallels? Yes, but not parallel straight lines. The Arctic Circle does *not* give the shortest distance between two of its points; if one had to go from one point on the Arctic Circle to the opposite point of that circle, it would be shorter to nip across over the North Pole.

Do not think that the surface of the globe gives a geometry entirely different from Euclid's. Most of the theorems of Euclid which do not depend on the idea of parallel lines remain true on the globe. For example, the theory of congruent triangles is just the same; the base angles of an isosceles triangle are equal; the perpendicular bisector of AB is the locus of points equidistant from A and B – these results hold just as well on the globe.

Two results which depend on parallels are 'The angles of a triangle add up to 180°' and Pythagoras' Theorem. Neither of these hold on the globe.

It is easy to give an example to show that the angles of a triangle do not add up to 180°. Consider the triangle formed by starting at the North Pole, going down the meridian of Greenwich until the equator is reached; there turn East and go a quarter of the way round the equator; you are now in longitude 90°E.; turn to the North and go straight back to the North Pole, along the meridian 90°E. This triangle has three angles, each of which is a right-angle. The sum is therefore 270°. On the globe, the sum of the angles of a triangle is not a fixed quantity. The larger the area of the triangle, the larger the sum of its angles.

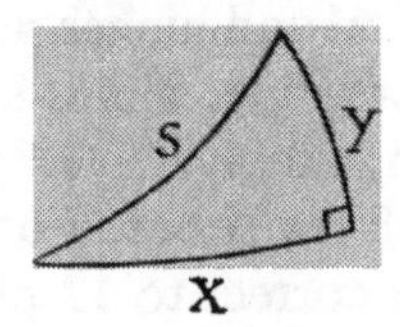

Figure 19

The same triangle shows quite clearly that Pythagoras' Theorem does not hold, for it is right-angled and yet its three sides are equal (Figure 19). If the radius of the globe is taken as the unit, the formula for the globe corresponding to Pythagoras is

$$\cos s = \cos x \cos y$$

The letters x, y, s here have the same meaning as in the formulae given earlier in this chapter, under the heading 'Number of Dimensions'. x and y are the distances along two 'straight lines' on the globe, at right angles to each other. s measures the third side of the triangle, the hypotenuse.

The geometry of the sphere evidently is very different from the geometry of the plane. Would it be possible for people to live in such a spherical universe for thousands of years, and to believe that they were living in a plane, until an Einstein arrived? It is possible to mistake a sphere for a plane; the Flat Earth controversy is evidence for that. The mistake is only possible so long as you are living on a very small part of the sphere. A very small piece of a sphere is almost indistinguishable from a very small part of a plane. An ant living in the Sahara desert may be pardoned if it believes in the Flat Earth theory.

It was mentioned above that the sum of the angles of a spherical triangle depends upon the area of the triangle. For a very small triangle, the area is almost zero, and the sum is almost 180°.

Pythagoras' Theorem also holds for very small triangles on a sphere. This may surprise you, as the formula above with cosines in it looks very different from $s^2 = x^2 + y^2$. But in fact, if x, y and s are small, the two formulae do agree. There is a series for the cosine, namely

$$\cos x = 1 - \tfrac{1}{2}x^2 + \tfrac{1}{24}x^4 - \tfrac{1}{720}x^6 \ldots$$

(*M.D.*, page 206). This series, incidentally, would give us a way of explaining what we meant by *cosine* to our angel friend; because the ordinary school explanation of cosine, based on drawing diagrams, would be impossible.

If x is a small number, x^4 and higher powers of x are very small compared with x^2. Accordingly, if x is small, $\cos x$ is, to a high degree of approximation, simply $1 - \frac{1}{2}x^2$; the higher powers of x hardly affect the sum of the series. For instance, if $x = 0{\cdot}001$, $1 - \frac{1}{2}x^2$ would give $\cos x$ correct to 12 places of decimals, which surely is enough. We use the same approximation for $\cos y$ and $\cos s$. Accordingly $\cos s = \cos x \cos y$ becomes

$$1 - \tfrac{1}{2}s^2 = (1 - \tfrac{1}{2}x^2)(1 - \tfrac{1}{2}y^2)$$

We multiply out the right-hand side of this equation, cancel 1 from each side, and find $\frac{1}{2}s^2 = \frac{1}{2}x^2 + \frac{1}{2}y^2 - \frac{1}{4}x^2y^2$. But x and y are both small, so x^2y^2 again only affects the twelfth place of decimals or thereabouts, and we neglect it. Multiplying what is left by 2, we have $s^2 = x^2 + y^2$, the familiar form of Pythagoras' Theorem.

We expected this answer, since we had observed that small parts of spheres resembled small parts of planes. The calculation served only to confirm what we already knew. Yet the calculation brought to our attention certain things about cosines which otherwise we might never have noticed. This is one of the ways in which mathematical discoveries are made. We know, from our experience, something to be true. We have formulae in which that truth can be stated. By trying to translate our intuitive knowledge into a formal proof, we discover new aspects of the formulae.

SPACE FINITE BUT UNBOUNDED

The traditional view of the universe is that it is infinite in size. The argument was, suppose you start to travel in any direction and keep on in a straight line. What is to stop you going on as long as you like? Surely we shall not find that the universe is bounded by a marble wall, that it is impossible to pass. If not, we can travel in a straight line as far as we like; that is to say, starting from the earth, the universe stretches out for an infinite distance in all directions. If this is not true, then there must be some sort of barrier surrounding us. And what about the other side of the barrier? Would not that be space too? Even if you admit the barrier, surely you admit something beyond it? In fact, space is still infinite, even if there are a few barriers scattered about.

It is sometimes held that relativity is a complicated theory. But was the old theory a simple one? It was a terrific idea, that of an infinite universe actually existing. Some philosophers have held that infinity can exist only as a possibility. There is no limit to the number of words that I *might* say, but the number of words that I *do* say is – mercifully – finite. Is there a contradiction in imagining an actually infinite universe? I can only say, I do not know.

With relativity, a new twist was given to the argument. The old argument was, either you must be able to go on for ever to new ground, or you must run into a barrier. But mathematicians have discovered a third possibility. It is possible for a universe to be finite – to occupy a limited amount of space – and yet for there to be no barriers. This is the idea that was at one time such a popular object of discussion – 'space finite but unbounded'.

The spherical universe described in the preceding section is in fact an example of this possibility. On the surface of the earth there are no barriers. If you walk in a 'straight line' on the earth, you can go on as long as you like (I assume you have some means for passing over seas, mountains, ice, etc.). But, all the same, the total surface of the earth is somewhat under 200,000,000 square miles. This illustrates the third possibility; if you walk long enough in a straight line, you may come back to where you started from!

You may think this is an absurd suggestion. But how would you prove that it was physically impossible? A straight line can be defined as a definite physical object, for our purpose – say as a thread stretched tightly between two points. Suppose you and a friend have an indefinite supply of string. You stretch this string, say, from the sun to the star Sirius. Then, all the while paying out string, you back away from each other, always holding the string tight and always having it pass close to Sirius and the sun, so that in effect you are producing the line joining these two stars. How do you know that, sooner or later, you will not bump into each other, back to back, just as you would if you backed away from each other round the equator? If you say, no, that is impossible, you are adopting a pre-scientific attitude; here is a proposed experiment in physics; you are claiming that you can predict the result of it without actually performing the experiment, and that with certainty (for 'impossible' expresses complete certainty). It is legitimate enough to say, 'If the universe is *as I imagine it*, then your idea is impossible'; this is much the same as saying, 'If the universe agrees with Euclid's geometry, your suggestion is impossible' – but we have seen there are other geometries, and the universe may agree with one of them. Certainly, the universe always seems to have a surprise up its sleeve. People have often thought they had the final truth about the world; but it has never turned out that way.

THE THREE-DIMENSIONAL SPHERICAL UNIVERSE

The surface of a globe has only two dimensions. Two numbers – latitude and longitude, for instance – are enough to fix your position. The idea that the universe might be represented as the surface of the globe is the form that general relativity might have taken among people who had previously believed themselves to be living in a Euclidean plane. Note that the idea of a globe is easy to us, but it would not be easy to the people themselves. We visualize a globe easily, because we live in three dimensions. But the people who actually live in a spherical universe have only two dimensions; the third dimension, and the globe embedded in it, are things which they could reach through an abstract mathematical argument, but not visualize or feel through their senses.

In the same way, if we want to construct a model of a 'finite but unbounded' 3-dimensional universe, we must be prepared to think in terms of 4 dimensions, and the thing which in 4 dimensions corresponds to a sphere. There are no mathematical difficulties in so doing.

It may help to start lower down. In the plane, which has 2 dimensions, every point is specified by 2 numbers, x, y. The points of the plane which are at unit distance from the origin form a circle. From Pythagoras' Theorem we see that, for such points, $x^2 + y^2 = 1$. We could imagine a universe consisting only of the points of this circle. It would be a universe of one dimension. In it, if you kept on walking, you would come back to where you started.

In 3 dimensions, every point is specified by x, y, z. The points at unit distance from the origin form a sphere. They satisfy the equation $x^2 + y^2 + z^2 = 1$, and give us a model for a universe of two dimensions.

So far, we have been able to visualize what was being discussed. But to the angel at the other end of the telephone all of this has been an abstract intellectual exercise. He is quite ready to take the next step, and cannot see why it bothers us.

In 4 dimensions every point is specified by four numbers, x, y, z, t. The distance of a point from the origin is given by $s^2 = x^2 + y^2 + z^2 + t^2$. All the points at unit distance from the origin satisfy the equation $x^2 + y^2 + z^2 + t^2 = 1$. They form a hypersphere (something like a sphere, only more so) and give us a model for a three-dimensional universe.

I think it will be clear that this specification is exact enough for mathematical treatment to be possible; the properties of such a universe can be worked out. As has been emphasized before, for the people in this universe only the points of the hypersphere have physical reality. The fact that they cannot get off it does not worry them: indeed, they cannot imagine such a thing at all. There is room in their universe for an earth and a solar system like ours, for *East, North* and *up*, all of which lie in the hypersphere itself. Going 'out of the hypersphere' means travelling at right angles to *East, North* and *up*; such an idea never troubles them.

I do not wish to go into the mathematical treatment, but simply to derive by analogy from the sphere two properties of this universe.

Let us go back to the universe which consisted simply of the surface of the globe. Light travels in a 'straight line'. On the globe we have seen that the equator and the meridians are 'straight lines'. Let us imagine a man standing at the North Pole of a spherical universe, and looking out at the rest of the world. What will he see? Nearest to himself he will see, of course, his immediate surroundings, ice and snow and polar bears. But his view will not terminate with a horizon. For light follows the contours of his globe. We should say it bent round the world. But he does not say that. If the light went straight in our sense, it would go clean out of his universe. The light goes as straight as it can without leaving the universe. Suppose he is facing in the direction of the meridian of Greenwich. He will be able to see Greenwich, and behind that parts of France and Spain and Africa (provided that these objects do not hide one another); beyond that he will see a great stretch of sea, the ice of Antarctica, and the South Pole. But his line of sight, still following the globe, will continue up the meridian 180° from Greenwich. Behind the South Pole he will see New Zealand, a few islands in the Pacific, the extreme tip of Siberia, and behind all this – the back of his own head! And in whatever direction he looks, the ultimate thing he sees, filling the whole horizon, is the back of his head.

In the three-dimensional analogy, the same thing would be true. If conditions were clear enough, and our telescopes were sufficiently powerful, in whatever direction we looked, we should see ourselves – or, more precisely, taking account of the time light would require to go right round the universe, whatever was on the earth some millions of years ago.

This is the first property I wished to explain. Now for the second. How could our man, living on the globe, discover that he was not inhabiting an infinite space? I suppose the weather to be too bad for him to see more than a mile or so; he cannot tell just by looking. We will suppose he has plenty of adhesive tape. He drives a post in, at the North Pole, and starts to wind adhesive tape on to it (Figure 20). Then he goes round and round, making the post ever larger. In time he will have covered the entire Northern Hemisphere with adhesive tape, and will be going round the equator. He does not notice any sudden change when he crosses the equator. The equator is 25,000 miles round; he does not notice that his circles are gradually growing smaller. He still thinks of himself as on the outside of a huge circle, with its centre at the North Pole, and continues applying adhesive tape.

But, as he approaches the South Pole, he will begin to notice a change. He is no longer on the outside of the circle, he is inside it. The more tape he puts on, the smaller his prison grows, and finally he has covered the whole earth with adhesive tape, and shut himself in at the South Pole.

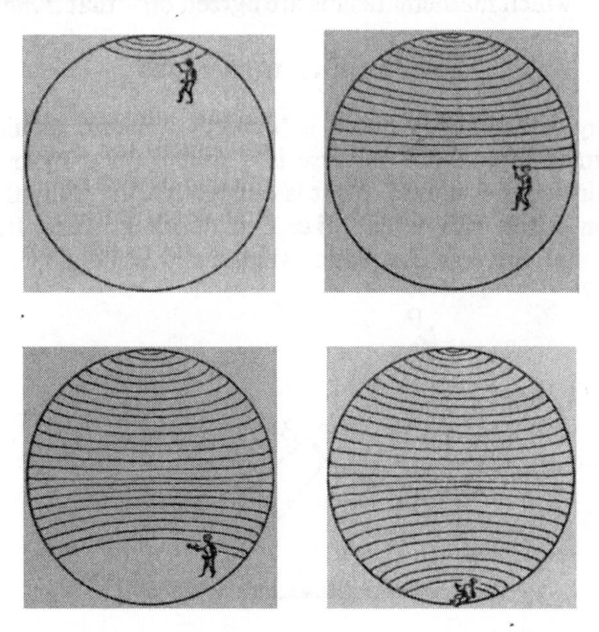

Figure 20

Now let us take the three-dimensional analogy of this process. Instead of enlarging a post with adhesive tape, we may think of ourselves starting with a cricket ball, and enlarging it by applying successive coats of paint. We have unlimited supplies of paint and we put more and more on. As the radius of the ball grows, naturally its surface seems less curved. But, in our hyperspherical universe, the time comes when we pass the equator. The surface of the enlarged ball has for some time appeared practically flat; we do not notice any sudden change. But as we continue we do notice something queer: the surface is not merely becoming flat, it is starting to curve the other way. We are no longer outside the paint; we are in a hollow sphere. As we continue to paint, this sphere becomes smaller and smaller, and we stop painting when we have no longer room to move. That is what it would feel like, if we lived in a space 'finite but unbounded'. And it has been seriously suggested that we do.

The experimental evidence is a matter for physicists. All that concerns a mathematician is the logical possibility of this theory, the fact – which mathematicians are agreed on – that *it may be so.*

EQUIVALENT UNIVERSES

You may feel unhappy about a 'curved' universe requiring an extra dimension. If the universe is a sphere (or a hypersphere) what is inside the sphere? What is outside it? But really the same objection arises with a 'flat' (i.e. Euclidean) universe. If a two-dimensional universe is a plane, what is above the plane, what is below it?

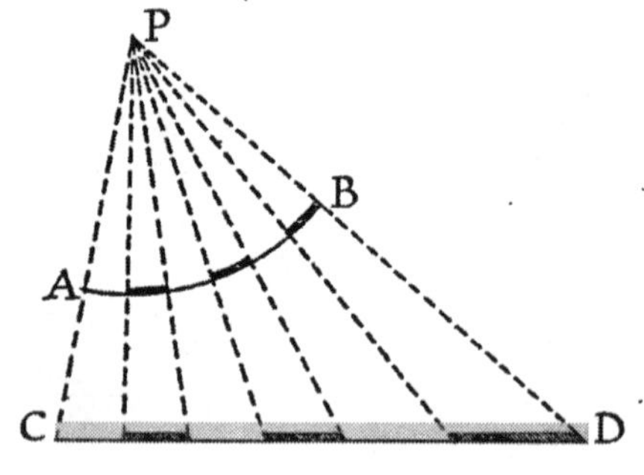

Figure 21

In fact, mathematically, there is no difference between the two ideas. Consider for example, people whose whole world is an arc of a circle, *AB* (Figure 21). Imagine a light put at any point *P*, and the shadows of the objects on the circle *AB* to fall on a straight line *CD*. Let us now imagine a universe created on *CD*, objects in which *behave exactly like the shadows cast from AB.* The two universes will then have exactly the same laws. It will be impossible for creatures living in *AB* to tell that they are not living in *CD*. These are in fact two different ways of describing the same universe. If we describe the universe as part of a circle, we can say that rigid objects keep a fixed length as they move. If we describe the universe as part of a straight line *CD* we shall have to introduce some effect (something like temperature) to explain why the length of an object changes as it moves along the line. But no experiment will enable one to decide in which universe one lives. The two theories are not different in their essential content; only in picturesque details. It is a matter purely of taste which picture one prefers.

To illustrate this point I will give you a 'flat' model of a two-dimensional universe embodying 'Possibility III'. This universe could also be got by considering objects sliding about on a suitable curved surface. The model here used is due to Poincaré (1854–1912).

POINCARÉ'S UNIVERSE

This universe is contained in the interior of a circle. At the centre of the circle the temperature is fairly high, but as you go away from the centre the temperature falls, and reaches the absolute zero at the circumference. The law for the temperature is quite a simple one. If a is the radius of the circle, at a distance r from the centre of the circle, the temperature is $T = a^2 - r^2$.

If now any object moves about in this universe, its size is affected by the variation in temperature. We shall suppose that the length of any object varies in proportion to the temperature T. At the circumference of the circle, where $r = a$ and $T = 0$, the length of the object will shrink to zero. The breadth varies in exactly the same way as the length.

But the inhabitants of this universe are not aware of the temperature. We suppose them to have no nerves sensitive to heat, so they do not feel temperature directly. Nor can they measure it by means of thermometers. The ordinary thermometer depends on the fact that mercury expands more rapidly than glass. But in this universe, every object expands and contracts in exactly the same way. If a creature is six feet long when it is at the centre of the circle, and it goes to a colder part, it will still find itself six feet long. If its length seems to us to be halved, so also is the length of the foot ruler with which it measures itself. It is still six times as long as the ruler.

But surely, someone may object, it will notice the fact that objects become of zero size on the boundary? The answer to this is that no one can reach the boundary (Figure 22). For as a creature starts to walk towards the boundary, its size (as seen by us) decreases; the nearer it comes to the boundary, the more rapidly it shrinks. To itself it seems to be taking steps of equal size; to us it seems to take shorter and shorter steps, the law being such that, however many steps it may take, it will never reach the boundary.[1] The boundary, from the creature's point of view, is infinitely distant. Lines which meet there are parallel. Thus, although this universe for us occupies a finite space, for the people in it, it is infinite.

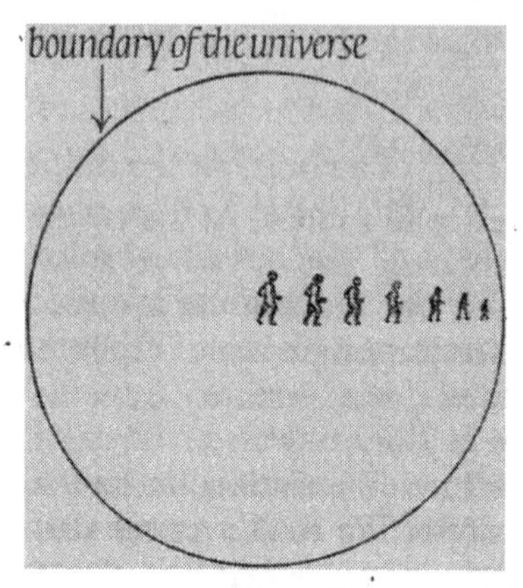

Figure 22

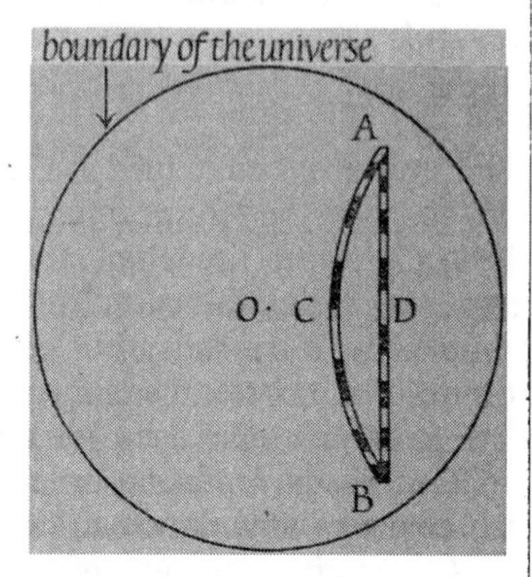

Figure 23

What will be a 'straight line' in such a universe? By definition, a 'straight line' is the shortest way from one point to another. The length of a path, for this purpose, is the length as measured by the creatures themselves. In Figure 23, *ADB* is a line that *we* should call straight. The light and dark segments of *ADB* show fifteen links of a chain connecting *A* and *B*. The links of this chain are – for the creatures – of equal length. The links near *D* are of course nearer the centre *O*; if these links were taken out to the

1. This can be seen by simple calculus. A length ds at the centre of the circle will, when transported to a distance r from the centre, occupy a space

$$dr = \frac{a^2-r^2}{a^2}ds.$$

So the creature regards a distance dr (in our measurement) as having a length

$$ds = \frac{a^2}{a^2-r^2}dr.$$

The creature's estimate of the distance from the centre to the edge of the circle is

$$\int_0^a \frac{a^2\,dr^2}{a^2-r^2},$$

which is infinite.

colder parts near *A* and *B* they would look just like the links at present near *A* and *B*.

Another chain is shown in the path *ACB*. The links of this chain are the same size as those of the first chain. But as *C* is in a warmer position than *D*, there are fewer links in the chain *ACB* than in the chain *ADC*. *ACB* is a 'shorter' path than *ADC*. As a matter of fact *ACB* is the 'straight line' joining *A* to *B*. Fewer links are needed for the chain *ACB* than for any alternative route.

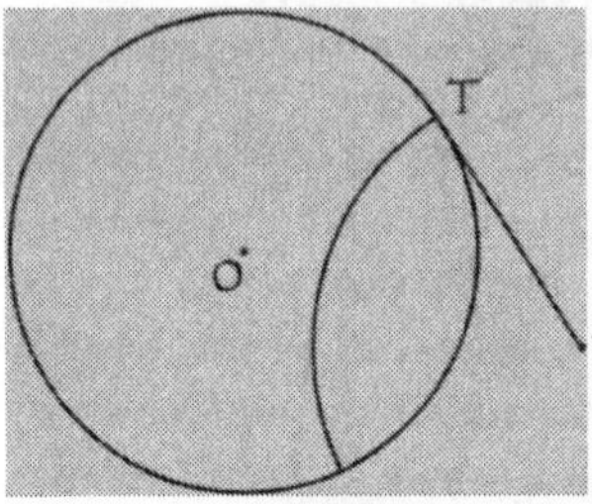

Figure 24

There is a simple geometrical construction for a 'straight line'. 'Straight lines' are in fact circles. Not any circle will do; it must be a circle that crosses the boundary circle at right angles (Figure 24). If *I* is any point outside the boundary circle, and *IT* is the tangent from *I* to the boundary circle, then the circle centre *I*, radius *IT* will do as a 'straight line'. One and only one such 'straight line' can be drawn to join any two points *A* and *B* in the universe. A stretched wire would naturally come to this shape; a ray of light would follow the same path. So, in the figure with the two chains, *ACB* is the shape that a chain stretched between *A* and *B* would take; the creatures could check this by eye; since light follows the track *ACB*, to a creature looking from *B* the point *C* would hide the point *A* – and this is the way we check that three points *A*, *C*, *B* are in line. So the shortest paths would look straight to the creatures inside the universe, although they do not look straight to us.

When the diagram illustrating Possibility III was drawn earlier, I remarked that it did not look right to us. Here is the same diagram with the same lettering, as it would be in Poincaré's universe. *AO* is the 'line' perpendicular to the 'line' *CD*. *P* is to the right of *O*, *Q* to the left. The dotted 'line' *NAM* is perpendicular to *OA*. As *P* moves to 'infinity' (i.e. to the boundary of the universe) the 'line' *AP* approaches the position *AR*.

This leaves a considerable angle between *AR* and *AM*. As *Q* moves to the left to 'infinity', the 'line' *AQ* approaches *AL*. Again, there is a considerable angle between *AL* and *AN*. If we take any direction from *A* that lies within the angle *LAN* or within the angle *RAM*, and proceed in a 'straight line' passing through *A* in this direction, we shall obtain a 'straight' path that does not meet *CD*. There are thus an infinity of 'lines' through *A* that do not meet *CD*.

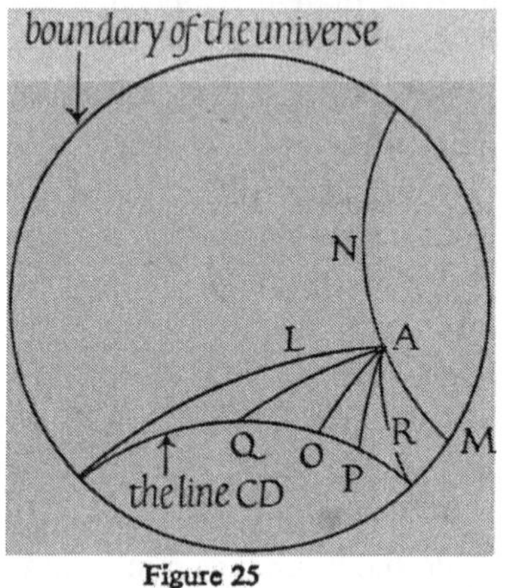

Figure 25

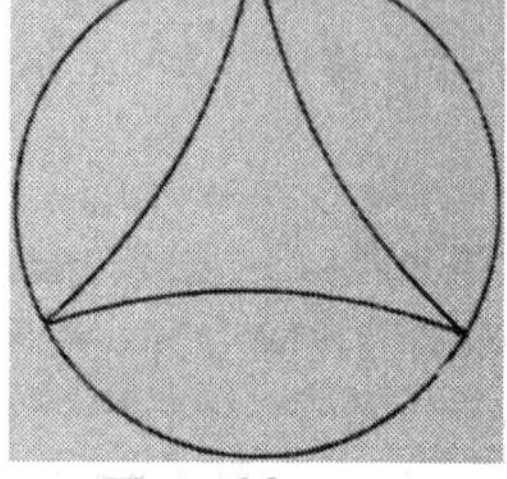
Figure 26

If we were to take a very small part of this universe, the temperature would vary very little within this small part. The geometry of a small part of this universe would accordingly be the same as Euclid's. The inhabitants of this universe, if their opportunities for travel were restricted, might well believe that they lived in a Euclidean plane. There is a moral in this for us, as we live within the solar system, which is a tiny speck in comparison with the distances between the stars.

In Poincaré's universe, the sum of the angles of a triangle is almost 180° for a small triangle, but the larger the triangle, the smaller the sum of the angles. If you look for instance at the triangle in Figure 25 formed by *AL*, *AR* and *CD*, the two angles on the boundary are nothing. The angle *LAR* is considerably less than 180°, and so the sum of the angles is less than 180°. In the triangle drawn in Figure 26 the sum of the angles is 0°.

By measuring the angles of a large triangle, it would be possible to discover whether one lived in a universe using Possibility I, II, or III.

CHAPTER SEVEN

Algebra without Arithmetic

Let U = the University, G = Greek, and P = Professor,
Then GP = Greek Professor.

Lewis Carroll

An old definition of mathematics is the study of number (arithmetic) and of shape (geometry). Number and shape are two very different ideas; why should these two be put together as one subject? It is true that shapes can be measured, and since Descartes brought in the use of graph paper it has been possible to turn every problem of geometry into one of algebra. But the idea of mathematics as arithmetic + geometry is much older than Descartes (1596–1650). In the traditional geometry, going back to the ancient Greeks, arithmetic entered hardly at all.

It seems to me that arithmetic and geometry, by a historical accident, were the first subjects to be given a fully logical form. Since then many other topics have been treated mathematically.

If one compares mathematics, biology, and art criticism, one gets some indication of the characteristics of mathematics. Art criticism does not – so far as I know – claim to be a deductive science. It does not start out from a clear definition of artistic value (I do not say beauty, because much great art is not beautiful) and assess pictures, symphonies, or books, on the basis of that definition. If our self-knowledge were sufficient, if we could predict with great exactitude our reactions to any object, a mathematical theory of art might conceivably be possible. I could imagine an anthropologist determining the precise mixture of aesthetic qualities that appealed to a primitive people, or the optimum mixture of sex, sadism, and sentimentality required by a modern film magnate. But in general art criticism is not, does not claim to be, and would not gain by being a deductive exercise.

Biology is partly deductive. There are certain general propositions admitted – such as that every animal requires some supply of energy to maintain life – and a system can to some extent be built up as to what kind of creature is capable of living in any particular environment. But many terms are undefined – notably the term 'living' – and along with the logical argument there is a constant appeal to observation, experience, and common sense. It is not a strict formal discipline.

In mathematical subjects a fully deductive treatment is aimed at. The ideas are supposed to be so clearly defined that one can develop them by a purely logical argument.

Sciences in general tend to become mathematical, since, with the development of a science, scientists gradually realize what they are talking about and become conscious of the methods by which they reason. As ideas become clarified, the possibility of logical development grows. There must however be some limitation to this process. If there were, for example, a mathematical theory of history enabling one to predict the future, the very knowledge so given would lead people to act differently.[1] A precisely formulated theory of art appreciation would have similar effects; it would cause people to change their tastes.

Mathematics, then, is concerned with reasoning about clearly specified things or ideas. There is no reason why mathematical symbols should stand only for numbers (as in arithmetic, algebra, trigonometry) or for points (as in geometry). They can stand for anything. Whatever they stand for, we develop them according to the properties that thing has.

AN ALGEBRA OF LOGIC

As an example of this, we will use symbols for the words 'and', 'or'. Since the use of these two words is well understood, our basic ideas have the necessary precision and clarity.[2]

We will let $+$ stand for *and*
$\cdot$ stand for *or*
a stand for 'Alfred is telling the truth'
b stand for 'Betty is telling the truth'
c stand for 'Charles is telling the truth'
$=$ stand for *has the same meaning as*

1. This idea is entertainingly developed in E. Hyams' novel, *The Astrologer*.

2. I do not know if any idea ever achieves complete precision. But all that matters for a formal theory, is that the idea is sufficiently precise for what you intend to do with it.

If any collection of these symbols is written, we judge the result to be true, false, or meaningless by using the vocabulary given above. We replace each symbol by the word or words it stands for, and pass judgement on the resulting sentence. What do you think of the three equations below? I shall answer this below the line of stars; if you want to work it out for yourself, do not read straight on.

(1) $$a + b = b + a$$
(2) $$a \cdot b = b \cdot a$$
(3) $$a \cdot (b + c) = a \cdot b + a \cdot c$$

* * * * *

Equation (1) reads, in effect: 'Alfred and Betty are telling the truth' has the same meaning as 'Betty and Alfred are telling the truth'. (Taken word for word it is slightly longer than this. 'Alfred is telling the truth and Betty is telling the truth', etc. The meaning however is as just given. I shorten the later examples in the same way.) Equation (1) is thus correct, though the truth it expresses is somewhat trivial.

Equation (2) reads, 'Alfred or Betty is telling the truth' has the same meaning as 'Betty or Alfred is telling the truth'. This again is a somewhat obvious truth.

Equation (3) would apply in the situation where witnesses were in conflict, Betty and Charles telling stories that corroborated each other, while contradicting the version told by Alfred. Equation (3) reads: 'Alfred is telling the truth or both Betty and Charles are' has the same meaning as 'Alfred or Betty is speaking the truth, and Alfred or Charles is speaking the truth'. This statement too is correct.

The remarkable thing is that equations (1), (2) and (3) are all equations that are correct in ordinary school algebra, with + meaning *plus* and $\cdot$ meaning *times*. The same pattern occurs in two widely different applications. The habits we have acquired for equations representing numbers can be carried over, and will give us logical truths. By multiplying out for example $(a + b)(c + d)$ and interpreting by the vocabulary above (d standing for the truthfulness of David) you can obtain a true, but lengthy, logical statement.

Another remarkable fact is that + and $\cdot$ can be switched around. If you like to take + as meaning *or* and $\cdot$ as meaning *and*, the algebraic procedure still gives correct logical results. In ordinary algebra one cannot switch + and . in this way. Accordingly, while all results of ordinary algebra are true in the algebra of logic[1] not all results of logical algebra are true in ordinary algebra. For example, the equation derived from (3) by switching + and .

(4) $$a + b \cdot c = (a + b) \cdot (a + c)$$

holds in logic – 'Alfred is truthful and Betty or Charles is truthful' means the same as 'Alfred and Betty or Alfred and Charles are truthful' – but does not hold for numbers.

But of course there is no need for every algebra to coincide with the algebra of ordinary numbers. So long as there are definite rules for manipulating the symbols, it does not matter what these rules are.

THE ALGEBRA OF CLASSES

Another algebra, also related to logic, is the algebra of classes. We might use, for example, s to stand for 'all small things' and g to stand for 'all green things'. We define sg to stand for 'all small green things'. This definition is very close to ordinary speech. The letter s expresses the property of being *small*; the letter g expresses being *green*. The two letters sg together combine just as the words *small* and *green* do in a sentence like 'It is a small green door'. But as algebraic symbols they look like a product; and in fact we shall find they behave like one.

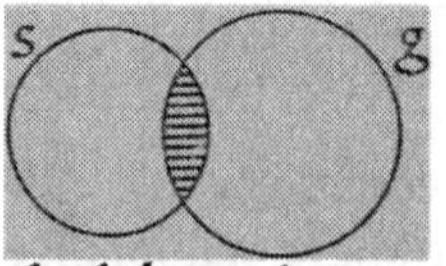

shaded area shows sg

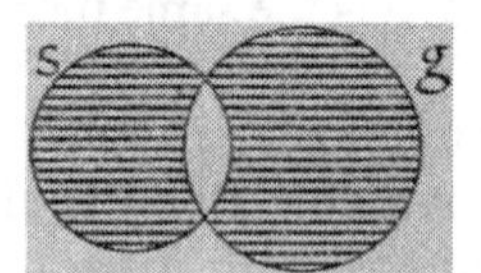

shaded area shows s+g

Figure 27

The 'sum' $s + g$ is defined in a rather strange way. It signifies all things that are small, or green, but not both! This is not a very natural definition, but it is the one that leads to the simplest rules for making calculations.

We may illustrate sg and $s + g$ by Figure 27. The circle

1. Of course, provided they have a logical interpretation. I have not given any interpretation of, for instance, $a - \frac{1}{2}b$.

labelled s is supposed to contain all the small objects in creation; that labelled g all the green objects.

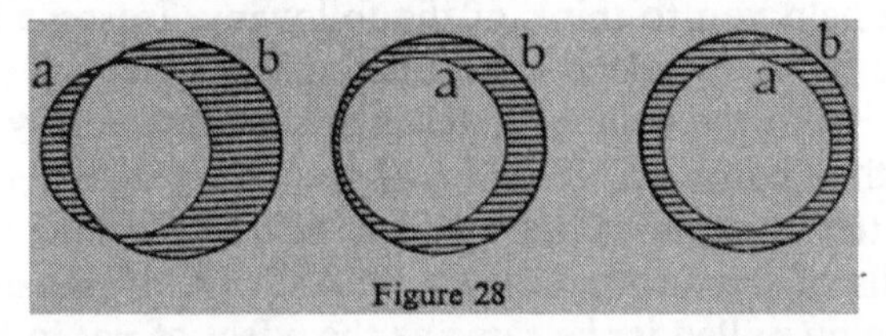

Figure 28

It is worth noticing what $a + b$ becomes if the class a lies entirely inside the class b. The diagrams below show $a + b$ when the class a slips into class b. (This can happen. Suppose a stands for 'all men under 60' and b for 'all persons born since 1900'. Figure 28 shows what happens between 1950 and 1970.)

You will see that, when a is part of b, $a + b$ means 'having the property b but not the property a'.[1] If it should happen that the class a grows until it coincides with b, the shaded area in the last diagram above will disappear. So $a + a$ stands for emptiness, for nothing. We can use the sign 0 for *nothing*. Thus $a + a = 0$. For *everything* we use 1. In Figure 29 the large rectangle is supposed to contain everything, the small circle to contain a.

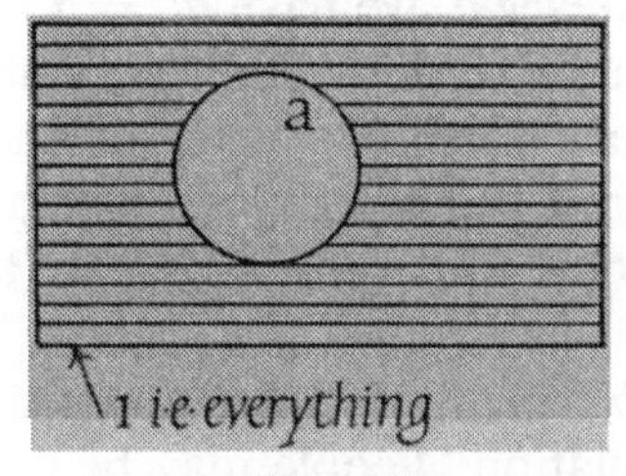

Figure 29

It will be seen that $1 + a$ means 'everything except a'. It stands for the property 'not a'.

You might think $1 - a$ would be a better sign for 'everything except a'. There is no harm in using this sign if you prefer it. Since $a + a = 0$, $+a = -a$. Plus and minus mean the same thing! On the whole, it is better to use $+$, because $s + g$ means the same as

1. This statement is a little loose. The class and the property, I suppose, ought not to be represented by the same letter a. But it does no harm, and saves space, as I use it here.

$g + s$, which is easier to remember than the same result with – instead of +. But you could use *minus* if you liked.

Of course $1 + 1 = 0$. This may strike you as strange. It may perhaps help you to think of the following. In some houses, for the sake of safety, electric switches are operated by long strings hanging from the ceiling. If the lights are off, and you pull the string, they come on. If you pull the string again, they go off. Pulling the string twice has the same effect as pulling it not at all. For pulls of the string, $1 + 1 = 0$. And indeed, the effect of the string being pulled is the same as the effect of *not* in logic. If the lights are *on*, and you pull, they become *not on*. If they are *off*, and you pull, they become *not off*.

In fact, we should expect to find $1 + 1 = 0$, to express this property of *not*. $a + 1$ means *not a*. $a + 1 + 1$ must mean *not not a*, or simply a. So $1 + 1$ should be 0. There would be something wrong if it was not.

We have discussed $a + a$. What about aa? This is very simple. Take g for 'green' as an example. gg means 'green, green things'. In poetry, to say something was 'green, green' might mean that it was very green, or 'Ah, me, how green it is!' or something like that. In logic it simply means 'all green things which are green', which is just the same as 'all green things'. So $gg = g$. And in the same way, whatever a, we have $aa = a$.

It is a most remarkable thing that the symbols so defined obey the laws of ordinary algebra – with, of course, the extra rule $1 + 1 = 0$ thrown in. I will not give a systematic proof that this is so, but will give some illustrations of how the algebra leads to correct results.

What will 'a or b' become in this symbolism? I use 'or' here in the sense of a regulation 'People may be admitted if they have paid at the gate or if they hold season tickets' – that is to say, not to exclude any eccentric or charitable person who may hold a season ticket *and* also have paid at the gate. If you look at the earlier diagrams, you will see that 'green or small (or both)' combines the shaded areas of the diagrams for sg and $s + g$. It is in fact $sg + s + g$. I do not know if it strikes you the same way, but as soon as I look at $sg + s + g$ I think 'That looks rather like $(s + 1)(g + 1)$.' As $(s + 1)(g + 1)$ multiplied out gives $sg + s + g + 1$, we have $sg + s + g = (s + 1)(g + 1) - 1$, and as -1 is the same as $+1$, this equals $(s + 1)(g + 1) + 1$.

But this last expression can be interpreted. $s + 1$ means 'not small'. $g + 1$ means 'not green'. The product $(s + 1)(g + 1)$ means 'not small and not green'. And the $+1$ at the end means 'not' or 'everything except'. So the whole expression means 'what remains when you remove everything that is not small and not green'. And this is what it should mean. For everything must fall in one of the four classes

(1) Small and green.
(2) Small but not green.
(3) Green but not small.
(4) Not small and not green.

And removing Class (4) does leave us exactly those things which are small, or green, or both. Not a very striking conclusion, but a verification that the ordinary laws of algebra (together with the extraordinary equation $+1 = -1$) do lead to correct logical conclusions.

Another example. Multiply out $a(1 + a)$. We have

$$a(1 + a) = a + aa = a + a = 0.$$

And this is correct, for $a(1 + a)$ means 'having the property a and also the property *not a*', which nothing has.

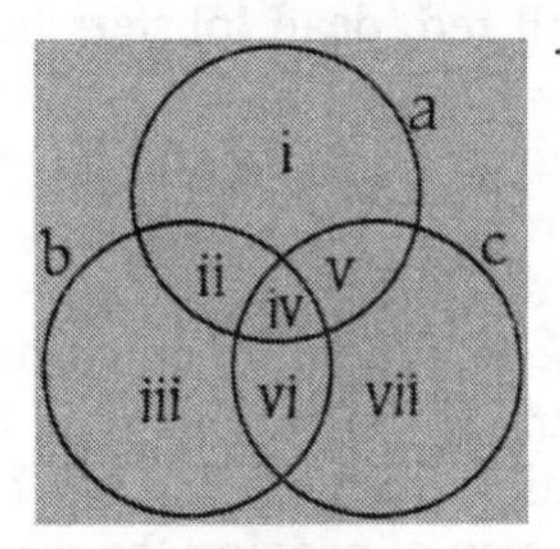

Figure 30

One can verify $a(b + c) = ab + ac$. We begin with $a(b + c)$. In Figure 30, $b + c$ covers the regions ii, iii, v, vii. So $a(b + c)$ contains regions ii and v. As for the other side of the equation, ab contains ii, iv; ac contains iv, v. $ab + ac$ consists of those regions which occur in ab or in ac but not in both; that is, ii and v.

The expression $a + b + c$ has some interest. It means those things that belong to *an odd number* of the classes a, b, c; that is, regions i, iii, iv, vii. You can verify this; and also that one arrives at the same result whether one regards $a + b + c$ as the result of adding a to $b + c$, or b to $a + c$, or c to $a + b$.

A final illustration: a puzzle asks, 'If all boiled, red lobsters are dead, and all boiled, dead lobsters are red, does it follow that all red, dead lobsters are boiled?' There are many ways of answering this question. I will deal with it by algebra.

As the whole question is about lobsters, we do not need to use a sign for 'lobsters'. We can work as if the universe contained nothing but lobsters, that is, regard lobsters as if they were 'everything'. Let b stand for 'boiled', r for 'red', d for 'dead'.

How are we to express in symbols that all boiled, red lobsters are dead? One way of looking at this statement is to re-word it, 'No boiled, red lobsters are not dead', i.e. the class of boiled, red, not-dead lobsters is empty. In symbols $br(1 + d) = 0$. If we multiply this out we get $br + brd = 0$, which means $br = -brd$, and as minus is the same as plus, this gives $br = brd$. This result we can check from its meaning; it says that the class of *boiled, red, dead* is the same as the class of *boiled, red.* And that is as it should be; we know that all lobsters with the properties *boiled, red* have also the property of being *dead.*

In the same way, all boiled, dead lobsters are red gives us the equation $bd = bdr$.

We are asked, are all red, dead lobsters boiled? That is, does the equation hold $rd = rdb$?

So the problem is; given $br = brd$ and $bd = brd$, does it follow that $rd = brd$? (The order of the letters b, r, d in products such as brd, bdr, rdb makes no difference.)

Now it does not *look* as if the third equation followed from the first two. But how can we be certain that there is no way of proving the third equation from the first two? After all, we might have overlooked some way of proving the result suggested.

If a proof existed, it would mean that all values of b, r, d that satisfied the first two equations also satisfied the third equation. For the third equation would hold whenever the first two did. But we can quickly show that this is not the case. For $r = 1$, $d = 1$, $b = 0$ makes the first two equations true, but turns the third equation into $1 = 0$, 'everything equals nothing', which is not true.

So it does not follow that all red, dead lobsters are boiled.

Algebra without Arithmetic

GROUPS OF MOVEMENTS

We now leave questions of formal logic, and consider a symbolism for representing the movements of bodies. In this section, by a *movement* I shall understand something that can be done to a rigid body, such as a rotation, or turning the body over, or transporting it. I exclude any kind of treatment which involves distorting the body, such as stretching, bending, or compressing it or any part of it. In the next chapter such distortions will not be excluded.

Capital letters will be used to stand for points of bodies, as is usual in geometry. Small letters will stand for movements, as for example we may say, let x stand for 'move one inch to the East' and t stand for 'turn through 90° about a fixed point O'. An equation such as $Q = xP$ will mean that Q is the point you get if you move P one inch to the East.

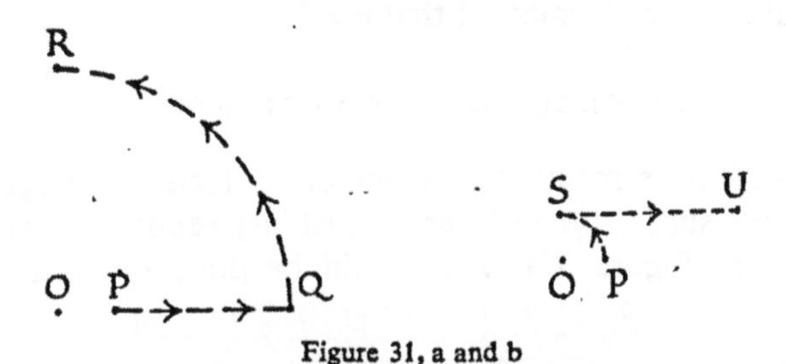

Figure 31, a and b

We can consider the effect of one movement followed by another. Suppose we take a point P, move it one inch to the East, getting Q, and then turn through 90° about O, which brings the point to the position R (Figure 31a). With symbols we can express this much more shortly, $Q = xP$, $R = tQ$. Since Q is xP, it is a natural thing to replace Q in the second equation by P, and write $R = txP$. This shows that R is obtained from P by applying *first* the operation x and *then* the operation t. The overall effect is denoted by tx. Notice that the operation performed second is written first. This is quite in accord with ordinary language. 'He left the burning house' means that the house first began to burn, and then he left it. (Some writers on groups use the opposite convention, so that there is considerable confusion caused by this small matter.)

The order in which the operations are carried out makes a difference. The diagram (Figure 31b) shows the meaning of xt. $S = tP$, $U = xS$, so $U = xtP$. U is not the same point as R. The effect of xt is therefore different from the effect of tx. So $xt \neq tx$. (This equation reads, xt is not equal to tx.)

In both the logical algebras we considered ab had the same meaning as ba. When $ab = ba$, we say multiplication is *commutative*. But when the order counts, we say multiplication is *non-commutative*. Very frequently multiplication is non-commutative. One needs to get used to distinguishing the order of symbols. A well-known example occurs in calculus,

$$x\frac{d}{dx}y \text{ is not the same thing as } \frac{d}{dx}xy.$$

A famous example is in quantum theory, where $pq - qp = h/2\pi i$. There is nothing mysterious in this equation. p and q are not numbers, they are operations, and the order in which they are carried out happens to matter; that is all.

THE GROUP OF THE RECTANGLE

Suppose we have a rectangle, a piece of cardboard perhaps, and we draw on a sheet of paper a space just large enough to contain the rectangle (Figure 32). This could be done by putting the

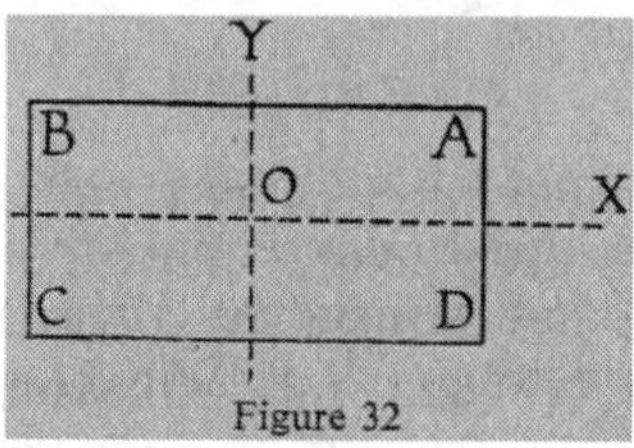

Figure 32

cardboard on the paper and running a pencil round the rim. Now there are various things I can do to the cardboard, which still leave it fitted exactly into its frame. (i) I could turn it over about the line OX, so that A and D change places, and B and C also. This operation we call p. (ii) I could turn it over about the line OY. Operation q. (iii) I could turn it through 180° about O, without lifting it from the paper. A and C would change places, as would B and D. Operation r. (iv) I could simply leave it alone as it is. Operation I.

If we blacken one corner of the rectangle so that it can be identified (we must blacken the cardboard on both sides) we have the diagrams (Figure 33) to show the effect of p, q, r and I.[1]

Figure 33

Now there are only four ways of putting the rectangle into its box. The black spot must come in one of the four corners, and once you have decided which corner, there is only one way of fitting the rectangle in. Accordingly, if we carry out two operations, one after the other, we must end with one of the four positions shown above. For example, suppose we want to know what qp is. We start with the spot in the North-East corner (as in all the diagrams above). p sends it to the South-East. Then applying q sends it to the South-West corner. But this is where the single operation r would send it. So $qp = r$. In the same way you can see that $pp = I$; this result is usually written $p^2 = I$. You can find the rest of the results easily. They are collected in the Multiplication Table below.

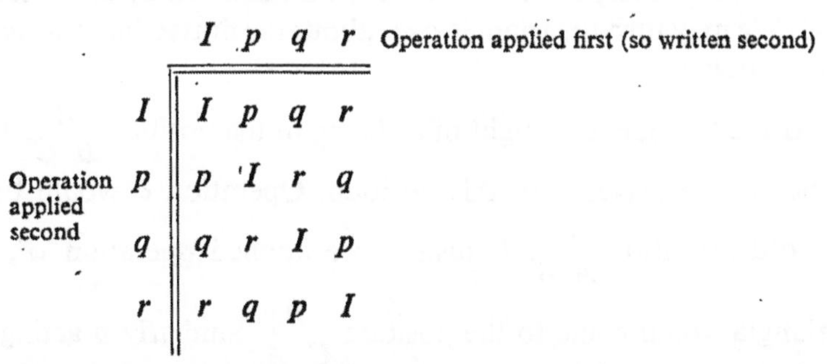

Operation applied second \ Operation applied first (so written second)	I	p	q	r
I	I	p	q	r
p	p	I	r	q
q	q	r	I	p
r	r	q	p	I

In giving a multiplication table for operations, one usually needs to indicate which operation is performed first, since in general ab and ba are different things. This has been done in the table above, though in the end it turns out to be unnecessary

1. The sign I is used partly because it is the initial letter of 'identical', partly because it looks like the numeral 1, with which it has many properties in common. To be consistent, a small letter should have been used for this operation; but i has associations with $\sqrt{-1}$. It seems best to be inconsistent.

in this particular case; e.g. pq and qp are both r. In fact, when any two of p, q, r are multiplied together, the result is the third. This multiplication table is commutative. But in general one should expect multiplication to be non-commutative, as in the example following.

THE GROUP OF THE EQUILATERAL TRIANGLE

Just as we made a frame for a rectangle, so we can take a piece of cardboard cut in the shape of an equilateral triangle, pencil round its outline on to a sheet of paper, and consider the different ways of fitting the cardboard into the pencil outline (Figure 34).

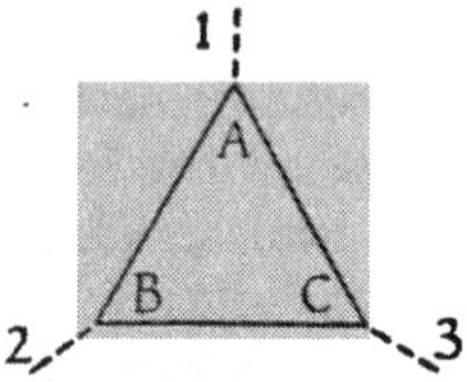

Figure 34

There are six possible movements. (i) We may leave the triangle as it is; operation I. (ii) We may rotate the triangle through 120° in the direction ↶; operation ω. (iii) We may rotate through 240°. This is the same thing as applying the operation ω twice, so it will be called operation ω^2. (iv) We may turn the triangle over, about the dotted line marked 1; operation p. (v) We may turn the triangle over, about the dotted line marked 2; operation q. (vi) We may turn the triangle over about the dotted line marked 3; operation r.

If the triangle is thought of as being in the position $\genfrac{}{}{0pt}{}{A}{B\ \ C}$, then operation I leaves it in this position. Operation ω would bring it to the position $\genfrac{}{}{0pt}{}{C}{A\ \ B}$. If instead we applied operation ω^2, the triangle would come to the position $\genfrac{}{}{0pt}{}{B}{C\ \ A}$. Similarly p acting on $\genfrac{}{}{0pt}{}{A}{B\ \ C}$ gives $\genfrac{}{}{0pt}{}{A}{C\ \ B}$; q acting on $\genfrac{}{}{0pt}{}{A}{B\ \ C}$ gives $\genfrac{}{}{0pt}{}{C}{B\ \ A}$; r acting on $\genfrac{}{}{0pt}{}{A}{B\ \ C}$ gives $\genfrac{}{}{0pt}{}{B}{A\ \ C}$.

Now there are only six ways of putting the three letters A, B, C on the corners of a given triangle, so the positions listed above must include all possible ways of putting the triangle into its frame. If we carry out two operations successively, for example if we do pr, this must bring $\begin{smallmatrix} & A & \\ B & & C \end{smallmatrix}$ to one of the six positions in the list; that is, it must have the same effect as a single operation, belonging to the six just listed. In fact pr on $\begin{smallmatrix} & A & \\ B & & C \end{smallmatrix}$ gives $\begin{smallmatrix} & B & \\ C & & A \end{smallmatrix}$. It has the same effect as ω^2. So $pr = \omega^2$. If we do the operations in the other order, that is, if we consider rp, a different result comes. rp changes $\begin{smallmatrix} & A & \\ B & & C \end{smallmatrix}$ to $\begin{smallmatrix} & C & \\ A & & B \end{smallmatrix}$, and thus has the same effect as ω. So $rp = \omega$. You may like to complete the multiplication table of this group before checking it against the table below.

Operation performed first (written second)

Operation performed second	I	ω	ω^2	p	q	r
I	I	ω	ω^2	p	q	r
ω	ω	ω^2	I	r	p	q
ω^2	ω^2	I	ω	q	r	p
p	p	q	r	I	ω	ω^2
q	q	r	p	ω^2	I	ω
r	r	p	q	ω	ω^2	I

OTHER GROUPS OF MOVEMENTS

Other groups of movements can be found by considering other figures. The more symmetrical the figure, the greater the number of movements in the group. One can consider as examples the letters of the alphabet. A cardboard *S* can be fitted into its outline in only two ways. Its group consists of the operation I, and of a rotation through 180° about the centre of the *S*. If we call the latter operation k, the multiplication table is simply

	I	k
I	I	k
k	k	I

Some letters have no symmetry at all, for example F, G, J, K, P, Q. Such letters, once placed in their frames, cannot be replaced in the frame by any operation except putting them back just as they were. The only operation in the group is I, and the multiplication table is trivial,

	I
I	I

There are many things you may notice about these multiplication tables. Groups are an important mathematical topic; they enter into an astonishingly wide range of mathematical enquiries. By means of group theory, you can obtain such different results as the insolubility of the equation of the fifth degree by algebraic methods, the fact that there are only 17 basically different wallpapers, and information on the structure of molecules in chemistry. At one time mathematicians felt groups were the key to the secret of the universe, and one can hardly blame them.

Apart from calling attention to the existence of group theory, this chapter tries to emphasize that mathematical symbols need not represent numbers, but may stand for words like 'or' or operations like 'rotate through 120°'. In Chapter 8 we shall consider matrices, which may be regarded as operations of a particular kind. They include the rigid movements we have considered above, and also other types of movement, in which rigidity is not preserved.

CHAPTER EIGHT

Matrix Algebra

> There would be many things to say about this theory of matrices, which, it seems to me, ought to come before the theory of determinants.
>
> *A. Cayley*, 1855

Descartes, as was mentioned earlier, showed that every geometrical result could be turned into an algebraic result. The points of a geometrical figure could be supposed drawn on graph paper. The position of each point would then be measured by a pair of numbers (x, y). Every property occurring in the geometrical theorem could then be translated into an algebraic relation between the xs and ys of the various points.

Some mathematicians admire Descartes' invention because it allows one to abolish geometry as a subject, and replace it by algebra. Others prefer to think in terms of geometry, without appealing to algebra. But the real value of Descartes' thought, it seems to me, is that it allows one to pass continually backwards and forwards between geometry and algebra. The meaning of an algebraic result can often be seen best by translating it into geometry; geometry gives a way of seeing and feeling algebraic abstractions. And geometrical results are often made more precise and clear when translated into an algebraic or arithmetical form.

Chapter 3 was called 'Algebra without Arithmetic'. In the latter part of that chapter we discussed geometrical notions, such as 'a rotation of 120°' or 'moving 3 inches to the East'. Even in these geometrical terms, numbers came, such as the 120 and the 3 just mentioned. We are now going to translate these geometrical ideas almost entirely into the language of number. We shall not need to say that an operation is a rotation; the numbers given will speak for themselves, and make it clear that the operation is a rotation. We shall in fact use four numbers to specify an operation. So arithmetic is coming back into the picture. But even so, the operations are what we want to discuss. One could describe a punch by saying that it felt like 2 tons of lead hitting you at 60 miles an hour. We could express this symbolically by saying that it was a (2, 60) punch, if we liked. But it would still be a punch; one would have to argue about it as a punch, not just as a pair of numbers. In the same way, in the remainder of this chapter, numbers will be used to specify operations, actions, movements. The correct way to handle these numbers is to think of the operations they represent. Of course, by thinking in this way, one can obtain rules for carrying out the calculations. The subject could be presented abstractly – for the benefit of angels on telephones – by simply stating these rules. In this way, it could be made purely a matter of arithmetic. This might have advantages from the viewpoint of logical exactness; but it would greatly hinder understanding and appreciation of the subject.

SPECIFYING OPERATORS

The operations considered at the end of Chapter 3 were of two types; either a body was turned over about a line, or it was rotated about a point. We are now going to investigate these operations in the spirit of Descartes.

In Chapter 3, an operation was represented by a small letter, so that it would appear different from the geometrical points *P*, *Q*, etc. In this chapter we shall have a lot of small letters representing numbers, *x*, *y*, etc., and also we shall have some capital letters, representing geometrical points. It is a pity that there are not more alphabets available for symbols. To avoid confusion both with numbers and points, I will represent operators by capital letters in heavy type, **A**, **B**, **C**, etc.

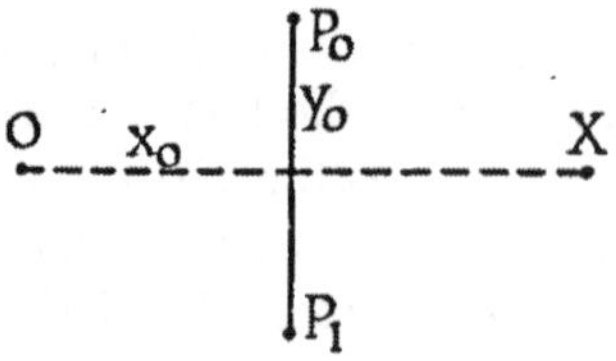

Figure 35

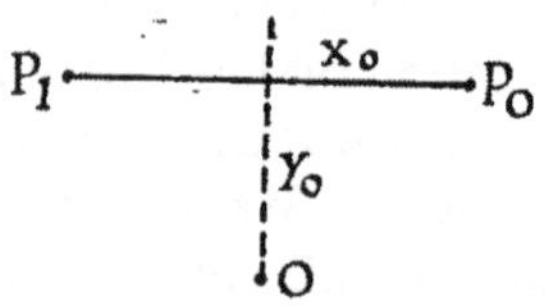

Figure 36

First consider operation **A** which means 'turn over about the axis *OX*'.

In the diagrams, we shall use P_0 to represent the position of a point before the operation is applied to it, P_1 the position after the operation.

(x_0, y_0) are the co-ordinates of P_0, (x_1, y_1) are the co-ordinates of P_1.

For operation A, it is evident from Figure 35 that P_1 has the same x-co-ordinate as P_0, but the sign of the y-co-ordinate is changed. So we have

$$\text{A}\begin{cases} x_1 = x_0 \\ y_1 = \qquad -y_0 \end{cases}$$

Now these two equations completely specify the operation A. They tell us what happens to any point (x_0, y_0).

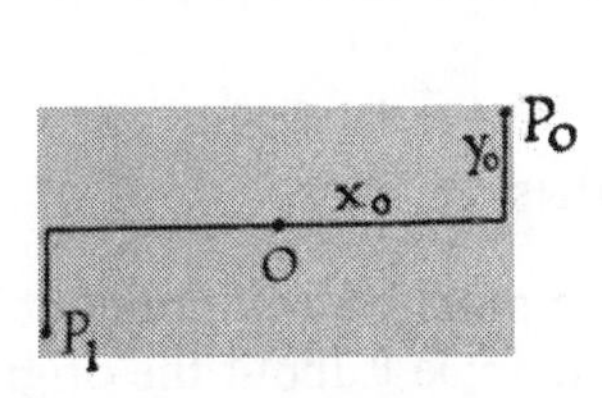

Figure 37

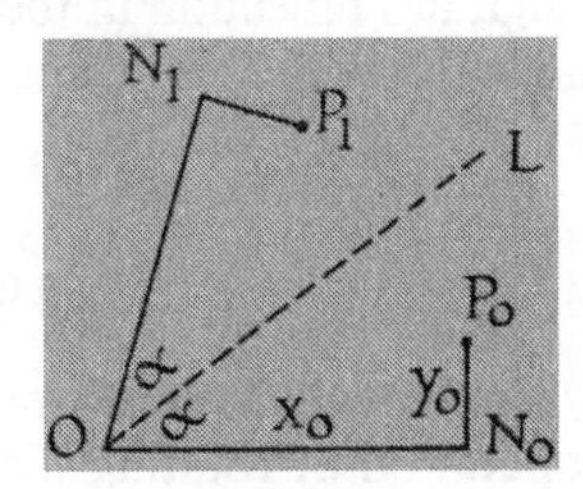

Figure 38

In the same way, if operation B means 'turn about OY', we have (Figure 36)

$$\text{B}\begin{cases} x_1 = -x_0 \\ y_1 = \qquad y_0 \end{cases}$$

Taking operation C to mean 'a rotation of 180° about O' we find (Figure 37)

$$\text{C}\begin{cases} x_1 = -x_0 \\ y_1 = \qquad -y_0 \end{cases}$$

These operations A, B, C are the ones that arose in the group of the rectangle in Chapter 7 and were there called p, q, r.

We may consider the operation D (Figure 38) for which, instead of turning about an axis, OX or OY, we turn about a line OL making an angle α with OX. We will call this 'reflection in the line OL'. This reflection sends P_0 to P_1. N_0 is the point on OX immediately under P_0. The reflection of N_0 is N_1. We know the length ON_1 equals ON_0, that is, x_0; and N_1P_1 equals N_0P_0, that is y_0. We know the direction of ON_1; it makes an angle 2α with OX. N_1P_1 is perpendicular to ON_1. Accordingly we can find the co-ordinates of P_1, by going from O to P_1 via N_1.[1] The result is

$$\mathbf{D}\begin{cases} x_1 = x_0 \cos 2\alpha + y_0 \sin 2\alpha \\ y_1 = x_0 \sin 2\alpha - y_0 \cos 2\alpha \end{cases}$$

There is a specially simple case of this, if $\alpha = 45°$. Then we have

$$\mathbf{E}\begin{cases} x_1 = y_0 \\ y_1 = x_0 \end{cases}$$

E signifies reflection in the line $y = x$. The effect of E is simply to interchange the co-ordinates of a point. Thus if P_0 is the point (3, 4), $P_1 = \mathbf{E}P_0$ will be (4, 3).

ROTATIONS

Very similar calculations enable us to specify a rotation. Let **F** be the operation of rotating through an angle θ about the origin O.

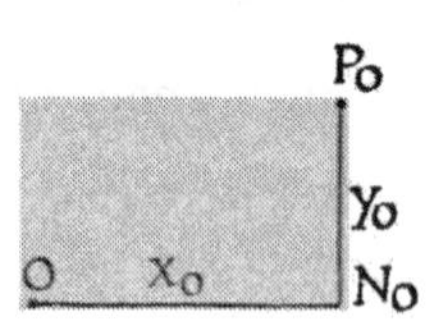

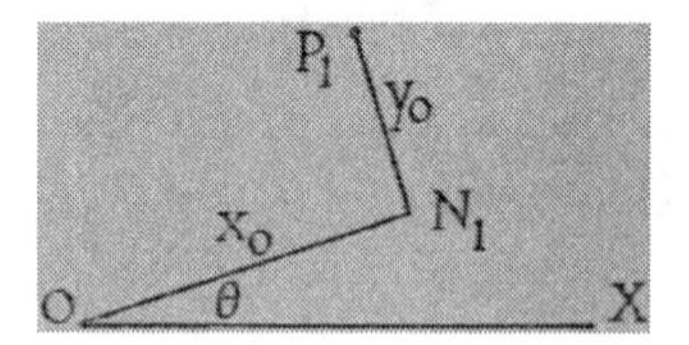

Figure 39a and b

Figure 39a shows the position of P_0. The lines of this diagram are then rotated through an angle θ, as in Figure 39b, to give the position of P_1. Following the route from O to P_1 via N_1 we find

$$\mathbf{F}\begin{cases} x_1 = x_0 \cos \theta - y_0 \sin \theta \\ y_1 = x_0 \sin \theta + y_0 \cos \theta \end{cases}$$

STRETCHES

One last example may be considered before we try to draw some general moral from these formulae. This operation will be called

1. Compare *M.D.*, Chapter 13.

stretching (Figure 40), G. This means that P_0 goes to a point P_1 on the line OP_0, the distance OP_1 being k times OP_0. k is a con-

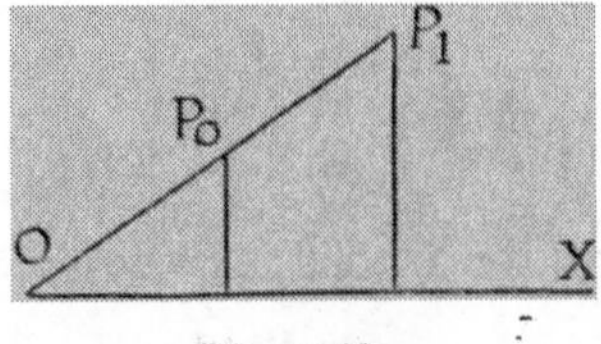

Figure 40

stant number. The effect of this operation is simply to change the scale of a diagram. By means of similar triangles the formula for it can be seen to be

$$G\begin{cases} x_1 = kx_0 \\ y_1 = \qquad\qquad ky_0 \end{cases}$$

MATRIX NOTATION

Looking back on all these results, we see that in each case x_1 and y_1 are given by linear expressions in x_0 and y_0. That is to say, in each case we have equations of the form

$$x_1 = ax_0 + by_0$$
$$y_1 = cx_0 + dy_0.$$

The operations can therefore be specified by saying what particular numbers a, b, c, d occur in each. When specifying them, it is customary to write the numbers a, b, c, d in the positions in which they occur in the equations, that is

$$\begin{pmatrix} a & b \\ c & d \end{pmatrix}$$

This is known as the matrix method for specifying operations. The set of numbers arranged in this form will be called *a matrix*.[1]

Each of the operations considered earlier can be written as a matrix. In some of the formulae, certain terms are missing. Thus,

1. The mould in which a printer casts type is called a matrix. Matrix in mathematics signifies simply that we have spaces into which numbers can be put.

for instance, in E we have only two terms on the right-hand side of the equations, instead of the four there could be. We regard the missing terms as having 0 in them. Thus the formula for E can be written

$$\mathbf{E}\begin{cases} x_1 = 0x_0 + y_0 \\ y_1 = x_0 + 0y_0 \end{cases}$$

In matrix notation we accordingly write

$$\mathbf{E} = \begin{pmatrix} 0 & 1 \\ 1 & 0 \end{pmatrix}$$

In the same way, we may write the other results; thus

$$\mathbf{A} = \begin{pmatrix} 1 & 0 \\ 0 & -1 \end{pmatrix} \quad \mathbf{B} = \begin{pmatrix} -1 & 0 \\ 0 & 1 \end{pmatrix} \quad \mathbf{C} = \begin{pmatrix} -1 & 0 \\ 0 & -1 \end{pmatrix}$$

$$\mathbf{D} = \begin{pmatrix} \cos 2\alpha & \sin 2\alpha \\ \sin 2\alpha & -\cos 2\alpha \end{pmatrix} \quad \mathbf{F} = \begin{pmatrix} \cos\theta & -\sin\theta \\ \sin\theta & \cos\theta \end{pmatrix}$$

$$\mathbf{G} = \begin{pmatrix} k & 0 \\ 0 & k \end{pmatrix}$$

Sometimes we want to show the symbols x_1, y_1, x_0, y_0 as well as the coefficients a, b, c, d. To show that a particular matrix is acting on a point with co-ordinates x_0, y_0 we write

$$\begin{pmatrix} x_0 \\ y_0 \end{pmatrix}$$

after the matrix. The full symbolism for the equation E given a little earlier would thus be

$$\begin{pmatrix} x_1 \\ y_1 \end{pmatrix} = \begin{pmatrix} 0 & 1 \\ 1 & 0 \end{pmatrix}\begin{pmatrix} x_0 \\ y_0 \end{pmatrix}$$

In this form there is not much economy in writing; nevertheless, this notation has its advantages, as will appear shortly. The most compact form of this statement is of course the original geometrical one, $P_1 = \mathbf{E}P_0$.

THE GENERAL MATRIX

The most general matrix appears in the equations

$$\begin{aligned} x_1 &= ax_0 + by_0 \\ y_1 &= cx_0 + dy_0 \end{aligned} \quad \text{or} \quad \begin{pmatrix} x_1 \\ y_1 \end{pmatrix} = \begin{pmatrix} a & b \\ c & d \end{pmatrix} \begin{pmatrix} x_0 \\ y_0 \end{pmatrix}.$$

We know that, for particular values of a, b, c, d this can represent a rotation, a reflection, or a stretch. But are these the only operations this can represent? What does the general matrix do to a geometrical figure? How can we visualize the effect of it?

Let us investigate this, for perfectly general values of a, b, c, d. We begin by seeing what it does to points on the axis OX. We take L_0, M_0, N_0 at distances 1, 2, 3 from O, and substitute their co-ordinates in the equations above.

O L_0 M_0 N_0 X

As L_0 is (1, 0), L_1 must be (a, c).
As M_0 is (2, 0), M_1 must be $(2a, 2c)$.
As N_0 is (3, 0), N_1 must be $(3a, 3c)$.

On plotting these points (Figure 41), we see they lie at equal distances along a straight line.

In the same way, by taking points P_0, Q_0, R_0 on OY, we find that (0, 1) goes to (b, d), (0, 2) goes to $(2b, 2d)$, (0, 3) goes to $(3b, 3d)$ (Figure 42).

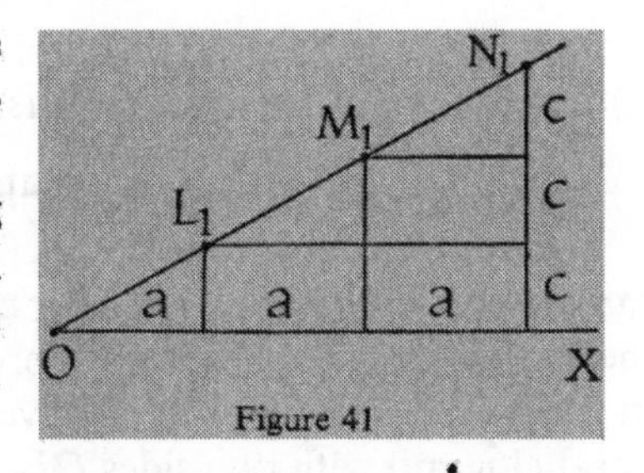

Figure 41

Thus

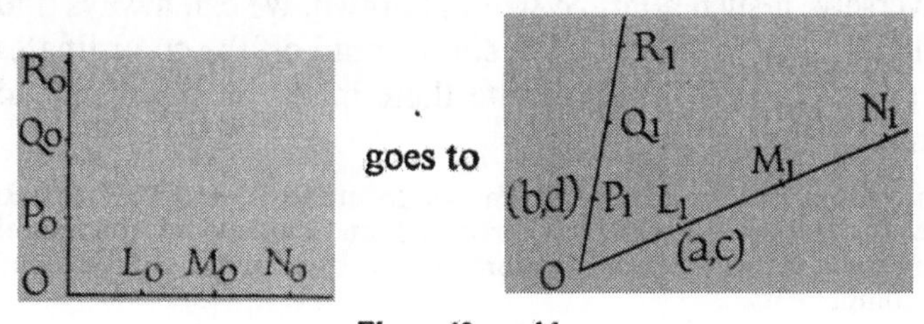

Figure 42a and b

Where does the point S_0 with co-ordinates (1, 1) go? By putting $x_0 = 1$, $y_0 = 1$ we find that S_1 is $(a + b, c + d)$. (Figure 43).

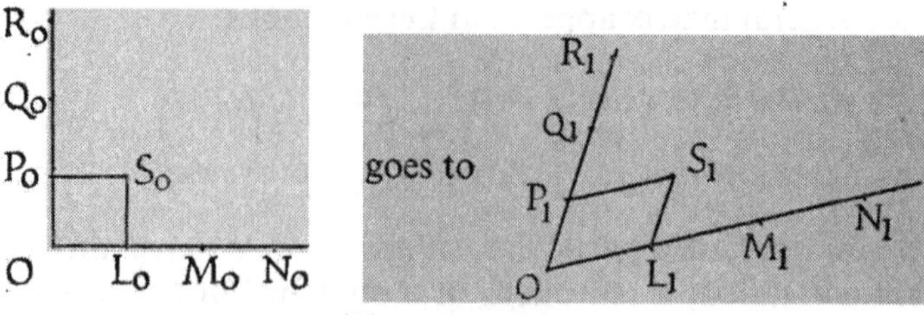

Figure 43a and b

We can show that $OL_1S_1P_1$ is a parallelogram. The corners of this figure have the co-ordinates (0, 0), (a, c), (b, d), $(a+b, c+d)$. It is a simple exercise to calculate the gradients of the sides, and to verify that opposite sides have equal gradients.[1]

In fact, as a little further investigation shows (Figure 44),

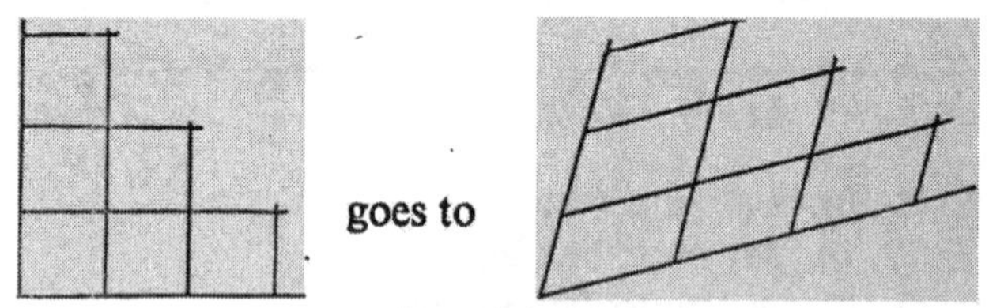

Figure 44a and b

Straight lines go to straight lines; parallelograms go to parallelograms; the origin stays where it is.

Accordingly, the effect of a matrix $\begin{pmatrix} a & b \\ c & d \end{pmatrix}$

can be seen without any need for calculation. The two numbers in the first column a, c give the co-ordinates of L_1. The two numbers in the second column b, d give the co-ordinates of P_1. The parallelogram with two sides OL_1 and OP_1 can be drawn, and by repeating it the whole figure is obtained.

Conversely, if such a lattice-work is drawn, we can always find a matrix that will produce it. We simply read off the co-ordinates of the points L_1 and P_1, and write these in the first and second columns of the matrix.

1. This result is of course well known to students of mechanics. It expresses the fact that the resultant force, as found by the parallelogram of forces, is the same as the resultant force found by adding horizontal and vertical components.

For example, to write down the matrix for a rotation through 90°. A rotation of 90° sends L_0 to the position (0, 1) and P_0 to the position (−1, 0). The two columns required are accordingly

$$\begin{matrix} 0 \\ 1 \end{matrix} \text{ and } \begin{matrix} -1. \\ 0 \end{matrix}$$

The matrix is

$$\begin{pmatrix} 0 & -1 \\ 1 & 0 \end{pmatrix}.$$

APPLICATIONS OF MATRICES

The way in which we have been led to the idea of matrix might suggest that matrices are only useful for the study of rotations, reflections, and other distortions of geometrical figures. This however is not the case; matrices pervade mathematics of almost every kind.

Some of the applications arise directly from the aspect of matrices we have been studying, that is, geometrical distortions. An obvious example is the theory of building construction. If a load rests on any material, whether it be india-rubber or steel, a change of shape is produced, a state of *strain*. The strain may be very small, but it is there. Even a steel bar cannot exert or transmit any force so long as it is exactly in its natural shape. (This is related to Hooke's Law. We can regard the molecules of the steel as if they were a great collection of little balls, held together by extremely strong springs.) The effect of the strain is that little squares of material become (to a sufficient degree of approximation) little parallelograms; the strain can accordingly be specified by means of a matrix.

In electricity and magnetism, a body in which electric and magnetic forces act behaves in many ways as if it were a material under strain. Here too matrices occur.

In aerodynamics – the study of air flowing past aeroplane wings – matrices also occur. If one imagines a small square drawn in smoke, which is pulled into the air stream, the shape of this square will change as it moves along. The way in which it changes will show what is happening in the stream. Matrices naturally are used to specify the process. In books on aerodynamics you will meet the term *irrotational flow*, which means that the little squares are distorted in shape, but do not rotate. (Some care is needed to make this concept precise.) The reference to rotation shows the link with what we have just been considering. The same term, irrotational, is also used in hydrodynamics (flow of liquids).

In all these examples there is a direct connexion with geometry. But there are many applications of matrices in which geometry does not provide a link. The link comes from the *algebraic form* of matrices. Matrices arose earlier in this chapter as a way of writing *linear equations*. Linear equations, naturally, can arise in almost any branch of mathematics. (In fact, to a large extent it is true that mathematical progress has been made only with problems which are in one sense or another linear. Non-linear problems offer very great difficulties to which, as yet, the answer is not known to any extent.) Where they do arise, matrices will be in demand.

Linear equations arise automatically in any problem which is concerned with *small quantities*. Suppose we have some function $f(x, y)$ and we are interested in what happens to this function in a small region near the origin, that is, near $x = 0$, $y = 0$. If $f(x, y)$ belongs to a very wide class of functions it will be possible to expand it in a series of powers of x and y, say

$$f(x,y) = k + ax + by + ex^2 + gxy + hy^2 + mx^3 + nx^2y + \dots$$

But x and y are small. So x^2, xy, y^2, etc., are very small indeed, and can be neglected. To this order of approximation,

$$f(x,y) = k + ax + by.$$

When $x = y = 0$, $f(x,y)$ is k. So $k = f(0,0)$. It follows

$$f(x,y) - f(0,0) = ax + by.$$

The expression on the left-hand side represents the *change* in f in going from (0, 0) to (x, y). On the right-hand side, we have $ax + by$, exactly the type of expression that led us to matrices earlier. But of course we need two such expressions to give a matrix. So in any problem concerned with the changes in two functions $f(x, y)$ and $g(x, y)$ – and there are many such problems – matrices will arise.

The small vibrations of any structure are an important engineering application of matrices. The matrices come in, as explained above, because the vibrations are *small*. Serious breakdowns can occur if in some machine the beat of an engine happens to coincide with the natural rate of vibration of some other part of the machine. (You have probably noticed in some motor-cars, at a certain speed some object in the car begins to chatter.) In certain conditions, the perpetual vibration can lead to a serious rupture. It is like the old story of soldiers breaking step when crossing a bridge. (As undergraduates some of us used to tap the Cambridge lamp-posts in the hope of finding their natural rate of vibration and causing them to disintegrate. We never succeeded.)

Quantum theory can be developed by means of matrices (Heisenberg, Dirac). The tensors of relativity are a generalization of matrices.

Matrices occur too in many branches of pure mathematics. Conic sections are an elementary example; projective geometry, groups, differential equations somewhat more advanced instances.

Matrix algebra is in fact one of the most striking examples of a pattern that arises in the most varied circumstances.

MULTIPLICATION OF MATRICES

As A and B both stand for geometrical operations, by **AB** we shall understand the matrix that represents the result of applying the operations **B** and A in succession.

If we consider any point P_0 – we have now finished with the diagrams in which P_0 stood for (0, 1) on the axis OY – if we consider any point P_0 operation **B** sends P_0 to P_1

$$P_1 = \mathbf{B} P_0$$

If now operation A is performed, it will send $\mathbf{P}_1$ to $\mathbf{P}_2$.

$$P_2 = \mathbf{A} P_1.$$

Combining these two results we write, much as in Chapter 3,

$$P_2 = \mathbf{A}\mathbf{B} P_0.$$

We will continue for the rest of this chapter to use the numbers 0, 1, 2 in this sense. A suffix 0 will go with the place where any object is first found; 1 will indicate an intermediate position after an operation has been performed; 2 will show where it has gone after two operations.

Taking the operations A and **B** to be those indicated by these symbols near the beginning of this chapter, we know from the group of the rectangle that **AB** = C.

This result can also be reached algebraically. $P_1 = \mathbf{B} P_0$ means

$$x_1 = -x_0$$
$$y_1 = \quad y_0$$

$P_2 = \mathbf{A} P_1$ means

$$x_2 = x_1$$
$$y_2 = \quad -y_1$$

From these equations it is child's play to deduce

$$x_2 = -x_0$$
$$y_2 = \quad -y_0$$

In matrix notation this would read:

$$\text{from} \quad \begin{pmatrix} x_2 \\ y_2 \end{pmatrix} = \begin{pmatrix} 1 & 0 \\ 0 & -1 \end{pmatrix} \begin{pmatrix} x_1 \\ y_1 \end{pmatrix}$$

$$\text{and} \quad \begin{pmatrix} x_1 \\ y_1 \end{pmatrix} = \begin{pmatrix} -1 & 0 \\ 0 & 1 \end{pmatrix} \begin{pmatrix} x_0 \\ y_0 \end{pmatrix}$$

$$\text{it follows that} \quad \begin{pmatrix} x_2 \\ y_2 \end{pmatrix} = \begin{pmatrix} -1 & 0 \\ 0 & -1 \end{pmatrix} \begin{pmatrix} x_0 \\ y_0 \end{pmatrix}.$$

But we could also combine the equations 'from ... and ... ' to give

$$\begin{pmatrix} x_2 \\ y_2 \end{pmatrix} = \begin{pmatrix} 1 & 0 \\ 0 & -1 \end{pmatrix} \begin{pmatrix} -1 & 0 \\ 0 & 1 \end{pmatrix} \begin{pmatrix} x_0 \\ y_0 \end{pmatrix}.$$

So we see that

$$\begin{pmatrix} 1 & 0 \\ 0 & -1 \end{pmatrix} \begin{pmatrix} -1 & 0 \\ 0 & 1 \end{pmatrix} = \begin{pmatrix} -1 & 0 \\ 0 & -1 \end{pmatrix}$$

that is, $\mathbf{A}\mathbf{B} = \mathbf{C}$.

The method used here is perfectly general. If we have any two matrices U and V, we can find UV as follows. We write down the equations $P_2 = \mathbf{U} P_1$, $P_1 = \mathbf{V} P_0$ in their algebraic form; that is to say, we have equations giving x_2 and y_2 in terms of x_1 and y_1, also equations giving x_1 and y_1 in terms of x_0 and y_0. Using the latter pair, we substitute for x_1 and y_1 in the former pair. This gives us x_2 and y_2 in terms of x_0 and y_0. The new equations will represent a relation $P_2 = WP_0$; the matrix W can be read off by looking at the coefficients in the equations found. Then $W = UV$.

I will now carry this through, and it will give us a rule for multiplying matrices. We will suppose

$$\text{U is } \begin{pmatrix} a & b \\ c & d \end{pmatrix} \text{ and V is } \begin{pmatrix} p & q \\ r & s \end{pmatrix}.$$

a, b, c, d, p, q, r, s are of course just numbers.

The equation $P_2 = UP_1$ written in full gives the pair of equations

$$x_2 = ax_1 + by_1$$
$$y_2 = cx_1 + dy_1$$

The equation $P_1 = VP_0$ written in full gives

$$x_1 = px_0 + qy_0$$
$$y_1 = rx_0 + sy_0$$

On substituting in the first pair the values of x_1, y_1 given by the second pair we find

$$x_2 = a(px_0 + qy_0) + b(rx_0 + sy_0) = (ap + br)x_0 + (aq + bs)y_0$$
$$y_2 = c(px_0 + qy_0) + d(rx_0 + sy_0) = (cp + dr)x_0 + (cq + ds)y_0$$

The numbers that occur as coefficients of x_0 and y_0 in the last expressions above give the numbers that should be inserted in the matrix W. Thus

$$\begin{pmatrix} a & b \\ c & d \end{pmatrix}\begin{pmatrix} p & q \\ r & s \end{pmatrix} = \begin{pmatrix} ap + br & aq + bs \\ cp + dr & cq + ds \end{pmatrix}$$

This could be regarded as a formula for multiplying two matrices. You could find the product of any two particular matrices by substituting the values of a, b, c, d, p, q, r, s. But in practice there is a better rule. If you look at $ap + br$ in the product, you will see that the letters a and b in it come from the top row of the first matrix. The letters p and r come from the left-hand column of the second matrix. If we disregard the rest of the result we see only

$$\begin{matrix} \rightarrow \\ \end{matrix}\begin{pmatrix} a & b \\ . & . \end{pmatrix}\begin{pmatrix} p & . \\ r & . \end{pmatrix} = \begin{matrix} \rightarrow \\ \end{matrix}\begin{pmatrix} ap + br & . \\ . & . \end{pmatrix}$$
$$\uparrow \qquad\qquad\qquad \uparrow$$

In the same way with all the other results; if we put an arrow against any row of the first matrix, and an arrow against any column of the second matrix, the numbers so singled out appear in the product matrix. The place where they appear is shown by putting two arrows at the side of the product matrix, in positions corresponding to the arrows on the left-hand side of the equation.

Thus the numbers from the top row and the right-hand column

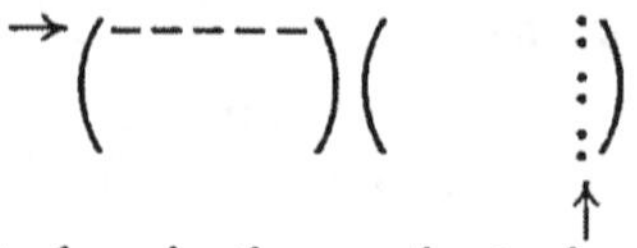

will appear at that place in the product where the top row meets the right-hand column

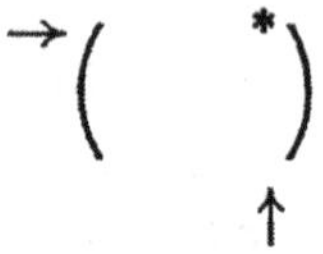

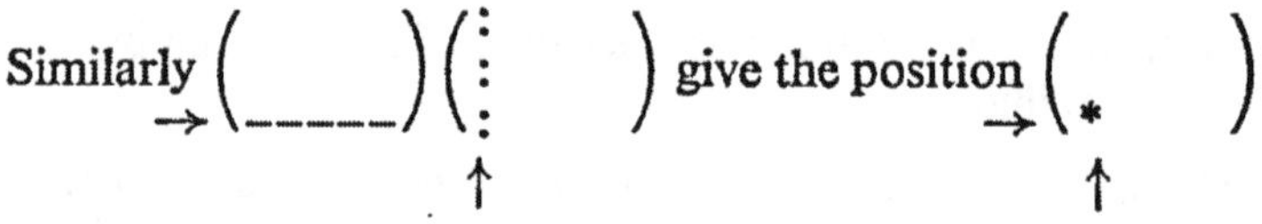

Similarly give the position

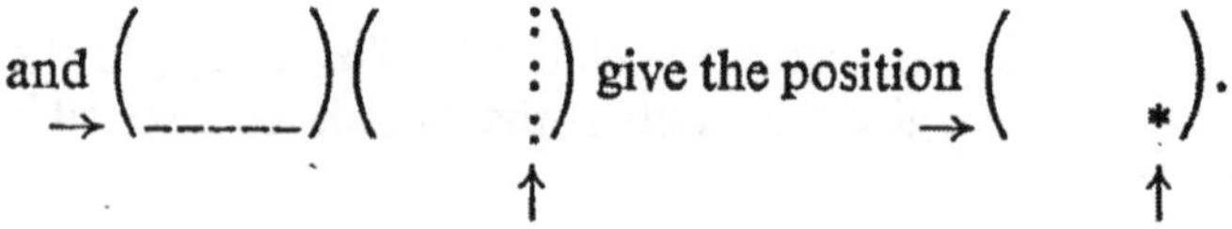

and give the position .

All of this is much easier to demonstrate on a blackboard than to write in a book.

I think you will see the rule for the number written in this place.

To take our first example, the numbers in the top row were *a*, *b*

and in the left-hand column *p*, *r*

Corresponding numbers are multiplied together; this gives *ap*, *br* and added, giving $ap + br$.

Anyone who works much with matrices becomes so familiar with this rule for multiplying that he could do it in his sleep. The finger of the left hand automatically moves across the rows of the first matrix, the finger of the right hand down the columns of the second. One multiplies the numbers while doing this, and then writes them down. The only pity is that nature has not given us three hands.

You will find this rule not difficult to become accustomed to, if you work a few examples. The best examples for this purpose are those where you know in advance what the answer should be; for then you will at once detect any slip you may make.

For example, the matrices **A**, **B**, **D**, **E** given earlier in this chapter all represented reflections. Now a reflection, done twice, lands you back where you were. Using $\mathbf{A}^2$ as usual for **AA**, we should find $\mathbf{A}^2$, $\mathbf{B}^2$, $\mathbf{D}^2$ and $\mathbf{E}^2$ all to be *I*, where *I* means what it did in Chapter 3, the operation of leaving things as they were. What is the matrix representing *I*?

Again, a rotation through the angle α followed by a rotation through β should give a rotation through $\alpha + \beta$. So, making use of the matrix **F** for rotation, we should find

$$\begin{pmatrix} \cos\alpha & -\sin\alpha \\ \sin\alpha & \cos\alpha \end{pmatrix}\begin{pmatrix} \cos\beta & -\sin\beta \\ \sin\beta & \cos\beta \end{pmatrix}$$

$$= \begin{pmatrix} \cos(\alpha+\beta) & -\sin(\alpha+\beta) \\ \sin(\alpha+\beta) & \cos(\alpha+\beta) \end{pmatrix}$$

This result, in fact, amounts to a way of proving the formulae for the sine and cosine of a sum.

You may like to verify that **FD** and **DF** are different. Each of them represents a reflection; the first about a line making the angle $\alpha + \frac{1}{2}\theta$, the second an angle $\alpha - \frac{1}{2}\theta$. The products therefore should have the form of **D**, but with α replaced by one of the angles just mentioned. The correctness of these statements can be seen geometrically.

ADDITION OF MATRICES

In algebra we usually carry out two operations; we add and we multiply. So far we have only discussed the multiplication of matrices. Is there any meaning which can be attached to addition of matrices?

We had no difficulty in defining multiplication. With operators, multiplication commonly means the result of carrying out the operations in turn. Applying this idea to matrices, we immediately obtained a definition of multiplication, and from that, the rule for multiplying matrices.

What advantages do we seek by carrying over such terms as *addition* and *multiplication* from arithmetic into other, very different subjects? We saw in Chapter 3 that certain problems of logic could be solved by introducing the signs + and . and using them exactly *as if* they represented addition and multiplication in ordinary arithmetic or algebra. We did not have to form new habits; our old habits gave us correct results. In formal mathematics we are only concerned with patterns. We are not concerned with what things are; we are only concerned with the patterns they make. If some idea enters into a pattern in exactly the same way as + enters into the pattern of arithmetic, no harm will come from calling that thing +.

How do addition and multiplication come into the pattern of arithmetic? What do we assume about the signs + and .? The most important things we assume are the following.

$$\text{(I)}\quad a + b = b + a$$
$$\text{(II)}\quad (a + b) + c = a + (b + c)$$
$$\text{(III)}\quad a.b = b.a$$
$$\text{(IV)}\quad (a.b).c = a.(b.c)$$
$$\text{(V)}\quad a.(b + c) = a.b + a.c$$

(I) is called the Commutative Law for Addition, (III) the Commutative Law for Multiplication, (II) the Associative Law for Addition, (IV) the Associative Law for Multiplication, (V) is called the Distributive Law.

Usually, when we leave ordinary arithmetic, we drop requirement (III). As mentioned earlier, we do not expect multiplication to be commutative; in fact, for matrices it usually is not.

A few simple assumptions have to be added if Subtraction and Division are to be brought in. I will not go into these here.

If one compares these assumptions with the axioms of Euclid's geometry – particularly allowing for the fact that Euclid uses a lot of terms without really explaining what they mean – you will see that the axioms of algebra are much simpler than those of geometry. The simplicity of algebra as compared with Euclid's geometry is due to this fact.

Any of the well-known formulae of algebra can be proved by

appeal to the axioms. It may interest you if I justify the formula $(a + b)^2 = a^2 + 2ab + b^2$ by means of the axioms. a^2 is an abbreviation for $a.a$ and $2ab$ for $a.b + a.b$, so what I really have to prove is $(a + b).(a + b) = a.a + (a.b + a.b) + b.b$.

By (V) $(a + b).(a + b) = (a + b).a + (a + b).b$. Now I cannot appeal again to (V) to multiply out $(a + b).a$, because (V) only tells me about $a.(b + c)$ with the multiplying *a in front of* the bracket that gets broken up into two parts. I must appeal to (III), which gives

$$
\begin{aligned}
(a + b).a + (a + b).b &= a.(a + b) + b.(a + b) \\
&= (a.a + a.b) + (b.a + b.b) \quad \text{using (V) twice} \\
&= a.a + (a.b + b.a) + b.b \quad \text{by an argument based on (II)} \\
&= a.a + (a.b + a.b) + b.b \text{ by (III);}
\end{aligned}
$$

what we wanted to show.

This may strike you as pedantic, and so it is *for the ordinary numbers*. But we intend to apply algebra to all kinds of operators, with which we are not so familiar as we are with arithmetic. We are by no means sure that the ordinary rules of algebra are going to apply. It is therefore most helpful to know that we can use our usual formulae if we can verify the five results (I) to (V) for our operators.

With matrices, for example, we cannot use

$$(\mathbf{A} + \mathbf{B})^2 = \mathbf{A}^2 + 2\mathbf{AB} + \mathbf{B}^2$$

because the proof of this formula twice uses Axiom (III), *which does not hold for matrices.*

We shall be entitled to bring the sign + into matrix calculations if we can find a way of defining it that will make axioms (I), (II) and (V) hold. Axiom (III) is going to be dropped anyway; we are building a non-commutative algebra. Axiom (IV) does not contain the sign +, so it is not affected by our definition of +.

Now obviously quite an investigation can start here, to find out what will be a suitable way of defining $\mathbf{U} + \mathbf{V}$ for matrices. Can a definition be found, that will give properties (I), (II), (V)? Is it the only definition possible to achieve this object? Are there several definitions that would do equally well? The latter questions I will not go into, but will only report that it has been found possible to make a satisfactory definition of + for matrices. This definition in fact is as simple as could be imagined. It is

$$\begin{pmatrix} a & b \\ c & d \end{pmatrix} + \begin{pmatrix} p & q \\ r & s \end{pmatrix} = \begin{pmatrix} a+p & b+q \\ c+r & d+s \end{pmatrix}$$

One simply adds together the numbers in corresponding positions. Thus, for example,

$$\begin{pmatrix} 1 & 2 \\ 3 & 4 \end{pmatrix} + \begin{pmatrix} 5 & 6 \\ 7 & 8 \end{pmatrix} = \begin{pmatrix} 1+5 & 2+6 \\ 3+7 & 4+8 \end{pmatrix} = \begin{pmatrix} 6 & 8 \\ 10 & 12 \end{pmatrix}.$$

It can be verified without tremendous labour that this definition satisfies all the axioms we expect it to. In regard to (V), since multiplication is not commutative, it is worth noting that it satisfies both

$$\mathbf{X}\cdot(\mathbf{U}+\mathbf{V}) = \mathbf{X}.\mathbf{U} + \mathbf{X}\cdot\mathbf{V} \text{ and } (\mathbf{U}+\mathbf{V})\cdot\mathbf{X} = \mathbf{U}\cdot\mathbf{X} + \mathbf{V}\cdot\mathbf{X}.$$

That is to say, the Distributive Law works all right, whether **X** is in front of or behind the sum U + V.

In matrix algebra, the simplest form to which we can bring $(\mathrm{U}+\mathrm{V})^2$, when it is multiplied out, is $\mathrm{U}^2 + \mathrm{U}\cdot\mathrm{V} + \mathrm{V}\cdot\mathrm{U} + \mathrm{V}^2$. If you look back at our earlier proof of the formula for the square of a sum, you will see how the lack of Axiom (III) stops any further progress.

HAS 3U + 4V A MEANING FOR MATRICES?

Quite early in school algebra, we meet expressions like $3x + 4y$. Algebra could not go very far without them. Can we, in matrix algebra, attach any meaning to 3U + 4V?

We certainly can. 2U is an abbreviation for U + U, and the meaning of U + U is already fixed by the definition of addition for matrices. In fact

$$\mathrm{U}+\mathrm{U} = \begin{pmatrix} a & b \\ c & d \end{pmatrix} + \begin{pmatrix} a & b \\ c & d \end{pmatrix} = \begin{pmatrix} 2a & 2b \\ 2c & 2d \end{pmatrix}.$$

So $$2\mathrm{U} = \begin{pmatrix} 2a & 2b \\ 2c & 2d \end{pmatrix}.$$

By adding a further U to this result, we find 3U (that is, U + U + U) to be

$$\begin{pmatrix} 3a & 3b \\ 3c & 3d \end{pmatrix}$$

Continuing thus, we see that for n any whole number,

$$n\mathrm{U} = \begin{pmatrix} na & nb \\ nc & nd \end{pmatrix}.$$

As usual, we take from a result just a little bit more than we are strictly entitled to. This formula has been justified for n a number like 2, 3, 4, 5, etc. But it suggests very strongly that we might still get quite a nice formalism if we assumed it to hold also for numbers like $1\frac{3}{8}$, -7, e, $\sqrt{2}$, π, and perhaps even sometimes numbers like $\sqrt{-1}$. And in fact it turns out that we do get a very satisfactory and quite simple algebra by taking this step. Accordingly we define, quite generally,

$$k\begin{pmatrix} a & b \\ c & d \end{pmatrix} = \begin{pmatrix} ka & kb \\ kc & kd \end{pmatrix}$$

This step is prompted by faith, but the consequences of taking it can be checked in the usual way by reason. It leads to a perfectly consistent and satisfactory pattern.

The things we can do with matrices now are (i) we can form a product UV of two matrices, (ii) we can form a sum U + V of two matrices, (iii) we can multiply a matrix U by a number k, so that kU is defined. It is worth noting that Uk is supposed to mean the same thing as kU.

ROTATIONS EXAMINED

With all this symbolism at our disposal, let us look back at the matrix we found for a rotation,

$$\mathbf{F} = \begin{pmatrix} \cos\theta & -\sin\theta \\ \sin\theta & \cos\theta \end{pmatrix}$$

Looking at this, it shows a certain pattern. $\cos\theta$ occurs in two places, $\sin\theta$ in two places. Let us separate these two parts. This we can do now, by using the sum of two matrices. It is easily seen that

$$\mathbf{F} = \begin{pmatrix} \cos\theta & 0 \\ 0 & \cos\theta \end{pmatrix} + \begin{pmatrix} 0 & -\sin\theta \\ \sin\theta & 0 \end{pmatrix}$$

The first of these contains the factor cos θ, which is a number. We can put this factor outside the matrix, since we know how to multiply a matrix and a number together. In the same way, we can take the factor sin θ out of the second factor. You can immediately verify that

$$\mathbf{F} = \cos\theta \begin{pmatrix} 1 & 0 \\ 0 & 1 \end{pmatrix} + \sin\theta \begin{pmatrix} 0 & -1 \\ 1 & 0 \end{pmatrix}$$

The matrix

$$\begin{pmatrix} 1 & 0 \\ 0 & 1 \end{pmatrix}$$

we have met before. It is the identity operator **I**, which leaves every point just where it was. What letter shall we use for the matrix next to sin θ? Let us call it **X** for a moment until we think of a better name. Suppose we multiply **X** by itself. We find by the usual rule for matrix multiplication

$$\mathbf{X}^2 = \begin{pmatrix} 0 & -1 \\ 1 & 0 \end{pmatrix}\begin{pmatrix} 0 & -1 \\ 1 & 0 \end{pmatrix} = \begin{pmatrix} -1 & 0 \\ 0 & -1 \end{pmatrix}$$

But the result here is what we should get if we multiplied the matrix **I** by the number −1. It could be written (−1) **I** or more shortly, −**I**. Thus we have the equation

$$\mathbf{X}^2 = -\mathbf{I}$$

As **I** plays the role for matrices that 1 plays for numbers, this suggests that we should think of **X** as being, in some sense, a square root of minus one. The appropriate name for it will accordingly be **i**.[1]

If we adopt this symbol instead of **X**, the equation for **F** above becomes

$$\mathbf{F} = \mathbf{I}\cos\theta + \mathbf{i}\sin\theta$$

1. Compare *M.D.*, Chapter 15.

which is strongly reminiscent of the very well-known expression $\cos\theta + i\sin\theta$.[1]

By following up this clue, it is possible to construct a theory of the complex numbers $a + ib$. We can *define* the symbol $a + ib$ as standing for the matrix $\mathbf{I}a + \mathbf{i}b$. This matrix written out in full is

$$\begin{pmatrix} a & -b \\ b & a \end{pmatrix}$$

It will then be found that these matrices, combined by the rules we have already discussed, behave exactly as we expect complex numbers to do.

But do not make the mistake of thinking that this approach is something entirely fresh. It is not. If you will look back in this chapter, just above the heading 'Applications of Matrices' you will find the very matrix

$$\begin{pmatrix} 0 & -1 \\ 1 & 0 \end{pmatrix}$$

that we have just used for $\mathbf{i}$. And this matrix there arose as standing for *the operation of rotating through* 90°, which is exactly how the operation $\mathbf{i}$ was explained first.[2] The matrix approach to $\mathbf{i}$ simply gives a new notation for the geometrical approach. Being algebraic in nature it may have advantages as allowing a more clear-cut development of the subject.

MATRICES IN GENERAL

Throughout this chapter we have considered only matrices containing 4 numbers arranged in two rows and two columns. This was done in order to keep explanations as simple as possible. But there is no reason for limiting ourselves to 'two by two' or 2 × 2 matrices, as they are called. We can just as easily write a set of equations in three variables

$$\begin{aligned} x_1 &= ax_0 + by_0 + cz_0 \\ y_1 &= dx_0 + ey_0 + fz_0 \\ z_1 &= gx_0 + hy_0 + kz_0 \end{aligned}$$

1. *M.D.*, Chapter 15, or any trigonometry textbook, De Moivre's theorem and related topics.
2. *M.D.* Chapter 15

in the matrix form

$$\begin{pmatrix} x_1 \\ y_1 \\ z_1 \end{pmatrix} = \begin{pmatrix} a & b & c \\ d & e & f \\ g & h & k \end{pmatrix} \begin{pmatrix} x_0 \\ y_0 \\ z_0 \end{pmatrix}$$

and this in turn in the still shorter form $P_1 = \mathbf{M}P_0$. These 3×3 matrices can be combined with each other by adding and multiplying, nor is there any great difference between their behaviour and that of 2×2 matrices. The rule for multiplying by rows and columns works in just the same way.

Square matrices can in fact be defined with any desired number of rows. $n \times n$ matrices behave like 2×2 matrices in practically everything except their geometrical interpretation (which of course requires space of n dimensions). 1×1 matrices are just ordinary numbers.

Rectangular matrices can be defined, with the number of rows different from the number of columns. A rectangular matrix, too, can be regarded as being the 'soul' of a system of equations. A practical application of rectangular matrices occurs in electrical circuit theory. Generally speaking, rectangular matrices are not as interesting mathematically as square ones.

CHAPTER NINE

Determinants

> But the next day one of his followers said to him, 'O Perfect One, why do you do this thing? For though we find joy in it, we know not the celestial reason nor the correspondency of it'.
>
> And Sabbah answered:
>
> 'I will tell you first what I do; I will tell you the reasons afterward.'
>
> Laurence Housman, *The Perfect One*

In almost any branch of mathematics, one finds a knowledge of determinants is required. Determinants have a claim to the attention of the applied mathematician because of their widespread usefulness; for the pure mathematician they represent a type of function with particularly simple and striking properties, obviously significant and deserving of study. The rules for calculating with determinants are simple; and many problems about determinants have striking and elegant solutions.

Perhaps the most remarkable thing about determinants is that, with all this evident mathematical significance and simplicity, their teaching is a real problem. If one looks at most textbooks dealing with determinants, the simple properties are reached by the most appalling calculations. Nor does this represent carelessness on the part of the authors. It is far from easy to find an elegant and illuminating way of presenting determinants.

In this chapter I intend to discuss the problem of teaching determinants. I shall first present the bare facts about determinants, the fairly simple rules they obey. If I were given two hours to teach a class of engineers how to use determinants, the bare facts, as outlined below, are about all I could cover. But the more intelligent students would certainly not be satisfied with these bare facts; I then consider the questions that would arise in these students' minds. Finally, I sketch various considerations in an attempt to illuminate the meaning of determinants. These are not, and do not claim to be, a rigorous exposition. They are simply intended to provide some sort of background; to satisfy a student that a treatment both rigorous and illuminating is possible; to make the properties of determinants seem natural.

THE BARE FACTS ABOUT DETERMINANTS

Suppose then I have my large class of engineers, and a very short period in which to show them how to use determinants. First of all, of course, I apologize to them for what scarcity of time is compelling me to do, to give the outer facts without the inner meaning. Then I proceed more or less as follows.

They may have seen in books a sign such as

$$\begin{vmatrix} a & b \\ c & d \end{vmatrix}$$

This is a kind of abbreviation. It stands for the number $ad-bc$. If the students knew anything about matrices or operators or anything of that kind, I would emphasize that it was *not* a matrix, *not* an operator, just a single number. (If a matrix were meant, the side lines would be curved not straight.) It stands for the single number $ad-bc$, nothing more, nothing less. For example,

$$\begin{vmatrix} 2 & 3 \\ 5 & 19 \end{vmatrix}$$

stands for the number 23, because $2 \times 19-3 \times 5$ is 23.

Again, the sign

$$\begin{vmatrix} a & b & c \\ d & e & f \\ g & h & k \end{vmatrix}$$

is also an abbreviation. It stands for the number

$$aek + bfg + cdh-ahf-bdk-ceg.$$

This also I would illustrate by a numerical example.

The square affairs with straight lines at the sides are called determinants. In a moment I will mention their main properties. Anyone with a knowledge of elementary algebra can verify these properties, by testing that the functions given above actually possess the properties stated.

(I) If two rows in a determinant are interchanged, the sign of the determinant changes. For example

$$\begin{vmatrix} c & d \\ a & b \end{vmatrix} = - \begin{vmatrix} a & b \\ c & d \end{vmatrix}$$

(II) If to the numbers in one row are added k times the numbers in another row, the value of the determinant is unaltered.

For example

$$\begin{vmatrix} (a+kc) & (b+kd) \\ c & d \end{vmatrix} = \begin{vmatrix} a & b \\ c & d \end{vmatrix}$$

(III) If rows and columns are interchanged, the value of the determinant is unaltered. This means that, for example, if instead of writing the letters a, b, c, d in alphabetical order across the rows, we write them in alphabetical order down the columns, it makes no difference.

$$\begin{vmatrix} a & b \\ c & d \end{vmatrix} = \begin{vmatrix} a & c \\ b & d \end{vmatrix}$$

Another way of saying this is that it makes no difference if we reflect the numbers of the determinant in the line running from the North-West to the South-East corner. This means that any statement that can truly be made about rows – in particular results (I) and (II) – can equally well be made about columns.

(IV) If all the numbers in any row are noughts, the value of the determinant is nought. For example,

$$\begin{vmatrix} a & b & c \\ 0 & 0 & 0 \\ g & h & k \end{vmatrix} = 0.$$

(V) Two determinants can be multiplied together by the following rule. (I give the rule only for 2×2 determinants here.)

$$\begin{vmatrix} a & b \\ c & d \end{vmatrix} \begin{vmatrix} p & q \\ r & s \end{vmatrix} = \begin{vmatrix} (ap+br) & (aq+bs) \\ (cp+dr) & (cq+ds) \end{vmatrix}$$

(VI) And finally I think I ought to point out to them that there is a connexion between 3×3 and 2×2 determinants. In fact

$$\begin{vmatrix} a & b & c \\ d & e & f \\ g & h & k \end{vmatrix} = a \begin{vmatrix} e & f \\ h & k \end{vmatrix} - b \begin{vmatrix} d & f \\ g & k \end{vmatrix} + c \begin{vmatrix} d & e \\ g & h \end{vmatrix}$$

The 2×2 determinant here that multiplies a is got from the 3×3 determinant by crossing out the row and column containing a. This leaves

$$\begin{matrix} e & f \\ h & k \end{matrix}$$

You will see that the other two determinants on the right-hand side of the equation are obtained in a similar manner, one of them by crossing out the row and column containing b, the other by crossing out the row and column containing c.

An extension of this rule gives us a way of defining 4 × 4 determinants. We use the equation

$$\begin{vmatrix} a & b & c & d \\ e & f & g & h \\ j & k & m & n \\ p & q & r & s \end{vmatrix}$$

$$= a\begin{vmatrix} f & g & h \\ k & m & n \\ q & r & s \end{vmatrix} - b\begin{vmatrix} e & g & h \\ j & m & n \\ p & r & s \end{vmatrix} + c\begin{vmatrix} e & f & h \\ j & k & n \\ p & q & s \end{vmatrix} - d\begin{vmatrix} e & f & g \\ j & k & m \\ p & q & r \end{vmatrix}$$

to define the 4 × 4 determinant on the left of the equation. The four 3 × 3 determinants on the right-hand side have a meaning which we know, as 3 × 3 determinants have already been defined.

I hope it is clear that one could repeat this procedure to obtain a definition of 5 × 5 determinants, and from them 6 × 6 determinants, and so on indefinitely. And at each stage you could verify – provided you were energetic enough to undertake the necessary algebraic calculations – that the properties (I), (II), (III), (IV), and (V) still applied to the larger determinants just brought in.

Six is not a large number of rules for an industrious student to learn. The rules are not particularly complicated. In two lecture periods of one hour each, with perhaps a bit of homework thrown in, it would probably be possible to give the students quite a fair idea of what could be done with determinants. One would have to work a number of examples on the board, and check the working of similar examples by the students themselves.

THE METHOD CRITICIZED

What criticisms would an intelligent student make of such an exposition?

The first criticism would probably be that the definitions seem arbitrary. I begin by defining 2 × 2 and 3 × 3 determinants, and in section (VI) I show how this definition can be extended to larger determinants. But how do I arrive at these definitions? Why do I regard it as natural to start with such definitions?

A second criticism would be as to the method of proof. I invite the students to check, by algebraic verification, that all the statements I have made about 2 × 2, 3 × 3, 4 × 4 and 5 × 5 determinants are true. But it is evident that I believe the properties

(I) to (V) are equally true of 6×6, 7×7 and in short $n \times n$ determinants. Whence do I derive this faith that determinants of whatever order, as found from my definition, will all enjoy the same properties?

A third criticism would be one of elegance. The results (I) to (V) are very simple. But the method I have suggested for verifying them involves long and formless calculations. Seeing the results are so simple, should there not be a correspondingly simple way of proving them? If I really understood why these results came, ought I not to be able to prove them by an argument containing little or no calculation?

If the instruction was being given in an unexpected emergency, the criticisms might be dismissed as untimely. But in any situation where reasonable leisure was available to the students the criticisms and the questions would be entirely justified.

Now I do not wish anything in the remainder of this chapter to be taken as a prescription for a perfect method of teaching determinants. Indeed I have the feeling that somewhere in the libraries of the world, if I had rather more time and somewhat better facilities for searching, I might find, all ready and complete, a far better exposition of this topic. What I write here should be regarded only as suggestions of the way in which one might seek for a better way of introducing to students the idea of determinants.

FROM MATRICES TO DETERMINANTS

The last diagram of the section 'The General Matrix' on page 110 shows the effect of a matrix operation, how it changes squares into parallelograms. Consider the effect of this operation on the area of any figure. If the figure initially consisted of a certain number of squares, it will after the operation consist of the same number of parallelograms. Fractions of a square will be changed into the same fractions of a parallelogram. That is to say, all areas will be changed in the same ratio. If the area of the parallelogram is g times the area of the square, then all areas will become g times the size they were before the operation.

If the matrix is

$$\begin{pmatrix} a & b \\ c & d \end{pmatrix}$$

what will g be? The corners of the parallelogram, it will be remembered, were the points (a, c), (b, d) and $(a + b, c + d)$, together with the origin O. Figure 45 shows this parallelogram enclosed in a rectangle. The lengths of various horizontal and vertical lines are marked. The area of the parallelogram can be found by subtracting from the area of the rectangle the various unshaded areas, namely, two rectangles of area bc each, two triangles each of area $\frac{1}{2}ac$, and two triangles each of area $\frac{1}{2}bd$. As the area of the large rectangle is $(a + b)(c + d)$ the area of the parallelogram is thus $(a + b)(c + d) - 2bc - ac - bd$, which simplifies to $ad - bc$. The square this parallelogram came from had corners (0, 0), (0, 1), (1, 0) and (1, 1). The area of the square was thus 1. Hence $g = ad - bc$.

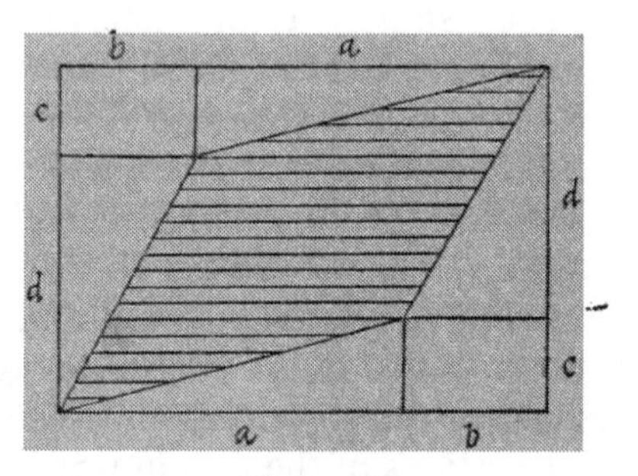

Figure 45

We recognize the answer. It is the determinant

$$\begin{vmatrix} a & b \\ c & d \end{vmatrix}.$$

We thus have an interpretation for a 2×2 determinant; *it represents the ratio in which the matrix changes areas.*

This idea does not work only for 2×2 determinants. If we go on to consider 3×3 matrices, we can find a similar way of seeing their effect on the points (x, y, z) of three dimensions. A cube will be changed by a matrix into a little box, every face of which is a parallelogram. The matrix will thus change the *volumes* of bodies in three dimensions in a constant ratio; this ratio will be the determinant of the matrix. We are not able to visualize the corresponding process in spaces of four or more dimensions; but logically no difference exists. The idea we have found applies in principle to all $n \times n$ matrices, whatever n.

The determinant then is a number associated with a matrix, and that is why we write them in such a way that each reminds us of the other.

When we use U as an abbreviation for a matrix, we shall use $|U|$ to stand for the determinant associated with that matrix; it is convenient to shorten this phrase to 'the determinant of the matrix'.

MULTIPLICATION OF DETERMINANTS

You may have noticed earlier in this chapter that Property (V) of determinants uses exactly the same rule as we had in Ch. 4 for multiplying matrices. If you look back to the section 'Multiplication of Matrices' in Ch. 4, you will find that the rule for multiplying arose very simply and naturally. It did not have to be dragged in from the blue; it arose naturally in the course of the working.

But it is now easy to see why the same rule allows us to find the product of two determinants. Suppose we have two matrices U and V. Let $|U| = g$, $|V| = h$. This means that the operation U multiplies every area by g, and the operation V multiplies every area by h. The product UV of the matrices means simply the result of applying first V then U. What will the determinant of UV be? That is to say, in what ratio does UV change areas? The answer is obvious. Operation V enlarges every area h times; then operation U multiplies the area g times. The combined effect will clearly be to enlarge areas gh times. So $|UV| = gh$. That is to say, if we multiply two matrices together, by the matrix multiplication rule, and take the determinant of the result, what we get will be the product of the determinants of the two matrices. This can be written

$$|UV| = |U| \cdot |V|$$

The importance of this result does not lie in its use for calculating the product of two determinants. If we know the value of one determinant is g and the value of a second determinant is h, then it is far simpler to multiply together the two numbers g and h than to appeal to rule (V). The rule is more likely to be used in the opposite direction. We have a complicated determinant to work out. We notice that the numbers in it have the form of the matrix product UV (that is, of the right-hand side of the equation given in rule (V)), and so we can reduce the problem to that of finding the two simpler determinants, $|\mathbf{U}|$ and $|\mathbf{V}|$. And we can also use it for arguments like those used later in this chapter to justify rules (I) and (II).

STRETCHES, ROTATIONS AND REFLECTIONS

In Chapter 8 we found matrices representing various geometrical operations. The effect of these operations on areas we know from geometrical considerations. If we work out the determinants, we obtain a verification of our argument, and also one unexpected result.

The determinant of the matrix **G** of Chapter 8 is

$$\begin{vmatrix} k & 0 \\ 0 & k \end{vmatrix} = k.k-0.0=k^2.$$

As **G** enlarges every length k times, it is quite correct that **G** enlarges areas k^2 times.[1]

Rotations carry areas round without changing them. We should expect 1 as the value of the determinant for a rotation. And in fact, for **C**, which represents a rotation through 180°, we have

$$|\mathbf{C}| = \begin{vmatrix} -1 & 0 \\ 0 & -1 \end{vmatrix} = 1.$$

For **F**, which represents a rotation through any angle θ, we have

$$|\mathbf{F}| = \begin{vmatrix} \cos\theta & -\sin\theta \\ \sin\theta & \cos\theta \end{vmatrix} = (\cos\theta)(\cos\theta)-(-\sin\theta)(\sin\theta)$$
$$= \cos^2\theta+\sin^2\theta = 1.$$

So this too verifies.

There remain the reflections **A**, **B**, **D**, **E**. For these we find

$$|\mathbf{A}| = \begin{vmatrix} 1 & 0 \\ 0 & -1 \end{vmatrix} = -1, \quad |\mathbf{D}| = \begin{vmatrix} \cos 2\alpha & \sin 2\alpha \\ \sin 2\alpha & -\cos 2\alpha \end{vmatrix} = -1,$$

$$|\mathbf{B}| = \begin{vmatrix} -1 & 0 \\ 0 & 1 \end{vmatrix} = -1, \quad |\mathbf{E}| = \begin{vmatrix} 0 & 1 \\ 1 & 0 \end{vmatrix} = -1,$$

after a simplification much like that done for $|\mathbf{F}|$ above.

1. Here and elsewhere, for the sake of having a definite image, I use a word which is strictly justified only in certain circumstances. **G** represents an *enlargement* only if k is larger than 1. If k is fractional, **G** represents a decrease in scale. And if negative values of k are in the reader's mind, I shall have to amend my statement even further. I find there is an economy of mental effort if one considers in the first place only the simplest, most easily visualized situation. Afterwards one can check whether or not the results apply in other situations.

Now here is something slightly unexpected. We normally think of turning over a piece of cardboard as leaving its area unchanged. In everyday life we think of area as always being positive. But when we start to find areas by calculation, an area may turn out to be negative. It is well known in calculus that, on finding an area by integration, the result may be negative. And indeed this fact is often useful. Again, in a well-known result of school mathematics, the area of a triangle is given by

$$\sqrt{s(s-a)(s-b)(s-c)}.$$

There is a square root here, and in extracting a square root, the question of $\pm$ always arises. So a triangle has two areas, one $+$ and one $-$, if we approach the question algebraically. The area of everyday life is simply the magnitude, the sign being neglected.

There are even in practical life occasions where the double sign of area can have some significance. There is a device known as an indicator which can be fixed to a steam engine. A pencil is connected to the pressure gauge, and a piece of paper to certain moving parts of the engine in such a way that the engine draws a kind of picture as it works. This picture takes the form of a closed curve, and the area inside the curve shows how much work the engine is doing at each stroke. But now suppose that, instead of allowing the engine to do work, we do work on it. We push against it in such a way that it retraces its steps; the pencil goes back along the curve it drew, but in the opposite direction. The area now represents not the work the engine has done but the work the engine has absorbed – *negative work*. So the same curve, traced in the opposite direction to the normal one, must be regarded as having negative area.

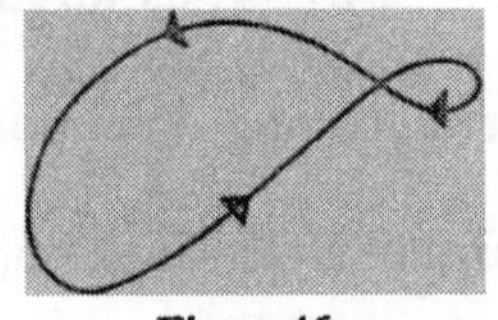

Figure 46

If one ever found an engine drawing a figure-of-eight curve like the one shown in Figure 46, it would mean that the engine was doing work in one part of its cycle – say the larger loop which it draws in an anti-clockwise direction – but absorbing energy in the smaller loop which it describes in the clockwise direction.

If you do not like the idea of negative areas, negative volumes and so forth, we could meet your views by rewording our earlier statement about the connexion of determinants and areas. We could say that the determinant of a matrix operation indicates two things; its magnitude indicates the ratio in which areas are changed; its sign indicates whether or not a reversal has taken place. For instance, in three dimensions, a matrix operation whose determinant was negative would change a car with right-hand drive into one with left-hand drive – that is to say, it would show the world as seen in a mirror.

But it is rather convenient to keep the idea of multiplying areas or volumes by -1. For example; we know that no amount of turning a car will change a right-hand into a left-hand drive; that is, it is impossible to combine any number of rotations to obtain a reflection. Can we prove this mathematically? Yes, in one sentence, if I am allowed to speak of $+$ and $-$ volumes. Every rotation multiplies volumes by $+1$; if several rotations are applied in succession, each rotation will multiply the volume by $+1$, and this can never lead to -1 times the original volume, which is what a reflection gives.

DERIVATION OF RULES (I) AND (II)

Now let us consider the following simple procedure. We have a 2×2 matrix operation **U**, with determinant g. We perform this operation **U** and then the operation **E**. What is the final effect of this on the areas of the plane? The operation **U** will multiply all areas by g. The operation **E**, which is a reflection in the line $y = x$, multiplies every area by -1, as we saw in the last section. The combined effect is to multiply areas by $(-1)g$, that is, $-g$. So the determinant of the combined operation, **EU**, must be $-g$. $|\mathbf{EU}| = -g$.

What is the actual form of **EU**? If we suppose $\mathbf{U} = \begin{pmatrix} a & b \\ c & d \end{pmatrix}$,

then, since $\mathbf{E} = \begin{pmatrix} 0 & 1 \\ 1 & 0 \end{pmatrix}$

$$\mathbf{EU} = \begin{pmatrix} 0 & 1 \\ 1 & 0 \end{pmatrix}\begin{pmatrix} a & b \\ c & d \end{pmatrix} = \begin{pmatrix} c & d \\ a & b \end{pmatrix}.$$

That is to say, EU is the matrix obtained from U by interchanging the rows. And its determinant, we saw above, is $-g$, that is to say, the original determinant with its sign changed. Thus we have proved Rule (I).

If you care to carry through a similar argument for the operation UE, you will find it gives the corresponding rule for the effect of interchanging columns.

Rule (II) can be obtained similarly, by considering the effect of successive operations.

Let the operation H be the matrix that changes the unit square into the parallelogram shown below. (Figure 47.)

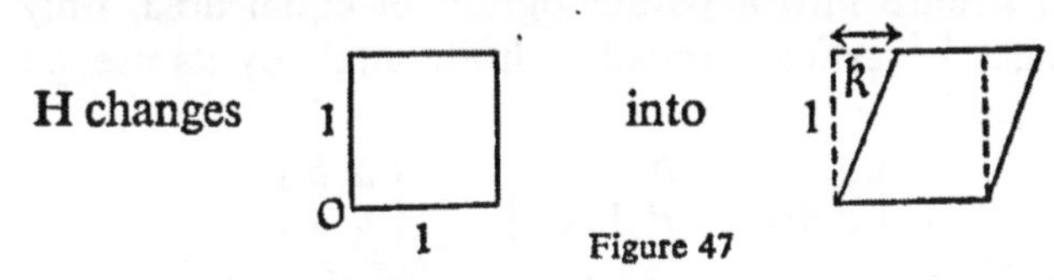

Figure 47

This operation does not involve any change in area; the triangle that the figure gains on the right just compensates for what it loses on the left. Nor does it involve a reflection. The change could be brought about gradually, by someone leaning with increasing force against the left-hand side of the square; a reflection can never be brought about gradually, within the plane. So the determinant of H must be +1. Hence

$$|\mathbf{HU}| = |\mathbf{H}| \cdot |\mathbf{U}| = 1 \cdot g = g.$$

The determinant of HU is the same as that of U.

What is HU? We shall first of all need to know the four numbers in the matrix H. This is easily done by the method explained at the end of the section 'The General Matrix' in Chapter 8. H sends the point (1, 0) to the position (1, 0), so the first column must be

$$\begin{matrix} 1 \\ 0 \end{matrix}$$

H sends (0, 1) to the position $(k, 1)$, so the second column must be

$$\begin{matrix} k \\ 1 \end{matrix}$$

(The information about where the points go is read off the geometrical diagram above.) So H must be

$$\begin{pmatrix} 1 & k \\ 0 & 1 \end{pmatrix}$$

and $$\mathrm{HU} = \begin{pmatrix} 1 & k \\ 0 & 1 \end{pmatrix}\begin{pmatrix} a & b \\ c & d \end{pmatrix} = \begin{pmatrix} a+kc & b+kd \\ c & d \end{pmatrix}$$

But we have seen that the determinant of HU equals that of U. And this is just what Rule (II) states.

The matrix $$\begin{pmatrix} 1 & 0 \\ k & 1 \end{pmatrix}$$

has a geometrical significance very similar to that of H: it also distorts a square into a parallelogram of equal area, only the displacement is vertical instead of horizontal. By its use we can prove

$$\begin{vmatrix} a & b \\ c+ka & d+kb \end{vmatrix} = \begin{vmatrix} a & b \\ c & d \end{vmatrix}$$

that is, that we may add k times the first row to the second.

Results for columns instead of rows may be obtained by considering products in the reverse order, UH for example.

GENERALIZATION OF THE METHOD

What shall we take for 3×3 matrices to correspond to the operation E? This may be seen most easily by considering the equations corresponding to $P_1 = \mathrm{E}P_0$, namely

$$\begin{aligned} x_1 &= \quad y_0 \\ y_1 &= x_0 \end{aligned}$$

These represent an exchange of two letters. The new x is the old y, the new y is the old x. If we are dealing with 3 variables x, y, z we use the same idea; we change round *any two* of the three letters. The third one of course remains unaltered. For example, if we change x and y around, and leave z the same, we have

$$\begin{aligned} x_1 &= \quad y_0 \\ y_1 &= x_0 \\ z_1 &= \qquad z_0 \end{aligned} \qquad \text{giving the matrix} \qquad \begin{pmatrix} 0 & 1 & 0 \\ 1 & 0 & 0 \\ 0 & 0 & 1 \end{pmatrix}$$

If instead we had changed y and z we would have arrived at the matrix

$$\begin{pmatrix} 1 & 0 & 0 \\ 0 & 0 & 1 \\ 0 & 1 & 0 \end{pmatrix}$$

or by changing x and z, the matrix

$$\begin{pmatrix} 0 & 0 & 1 \\ 0 & 1 & 0 \\ 1 & 0 & 0 \end{pmatrix}.$$

If you put any one of these matrices in front of a 3×3 matrix and multiply, you will find two rows are interchanged as a result.

Geometrically, each of these matrices has the effect of a reflection in a plane mirror. Taking x, y, z as *East, North, up*, the matrix first found here corresponds to a reflection in a vertical mirror, standing on a line in the North-East direction. This is shown in Figure 48.

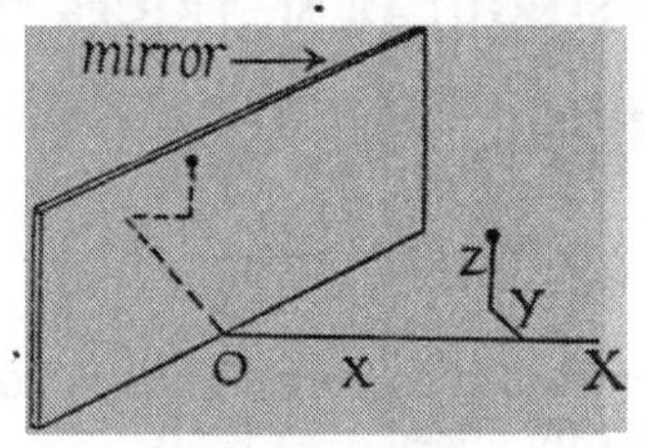

Figure 48

In the same way we can find a matrix in three dimensions to correspond to H. The matrix in fact is

$$\begin{pmatrix} 1 & k & 0 \\ 0 & 1 & 0 \\ 0 & 0 & 1 \end{pmatrix}$$

It has the effect of pushing a cube somewhat out of the straight without changing the volume. Something like a slice of cheese is lost on one side and gained on the other.

The method in fact can be carried over to $n \times n$ matrices. The difficulty that would have to be overcome in finding a strictly rigorous exposition would be the preliminaries needed to explain exactly what 'volume' meant in space of n dimensions. The details of this we will not go into here. It is sufficient if it has been shown that determinants do have a simple geometrical significance, and that their properties can be seen, at any rate for 2×2 and 3×3 determinants, by means of simple geometrical arguments.

It may be worth pointing out one thing which this method does not do; it does not establish Property (III) of determinants. In fact Property (III) is quite different from Properties (I) and (II). It is impossible to find matrices that will do for Property (III) what **E** and **H** do for (I) and (II). That is to say, it is impossible to find matrices, made out of constant numbers, that by multiplication will turn

$$\begin{pmatrix} a & b \\ c & d \end{pmatrix} \text{ into } \begin{pmatrix} a & c \\ b & d \end{pmatrix}.$$

Just how to fill this gap will not be discussed here.

SINGULAR MATRICES

If we look at the matrix

$$\mathbf{M} = \begin{pmatrix} 1 & 1 \\ 2 & 2 \end{pmatrix}$$

we see that its determinant is zero. That is to say, the effect of applying it is to multiply every area by 0.

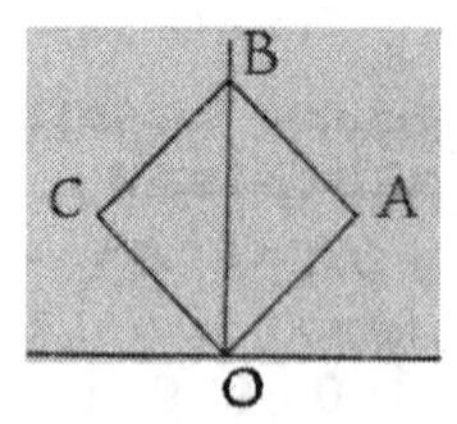

Figure 49

Let us consider what it does to the square $OABC$ in Figure 49, O being the origin, A the point (1, 1), B the point (0, 2) and C the point (–1, 1).

The equations corresponding to **M** are

$$x_1 = x_0 + y_0$$
$$y_1 = 2x_0 + 2y_0$$

By substituting the co-ordinates given above for (x_0, y_0) in these equations we find that the origin (0, 0) goes to (0, 0); the point A, (1, 1) goes to (2, 4); the point B, (0, 2) goes to (2, 4); the point C, (−1, 1) goes to (0, 0). So O and C both go to the position (0, 0); A and B both go to the position (2, 4).

In other words, the matrix squashes the sides OC and AB of the square down to nothing at all. It is by doing this that it manages to reduce the area of the square to zero. In fact every area is reduced to zero. You will find that whatever values you may give to (x_0, y_0), the point (x_1, y_1) will always lie on the line $y = 2x$. The matrix thus compresses the whole plane into a line.

A matrix that has determinant zero is called *singular*. A singular matrix always pushes some point, like C in the example above, into the origin.

The two conditions are in fact equivalent. If some point, that to begin with is distinct from the origin, is pushed into the origin by a matrix, then the determinant of that matrix is zero. And if the determinant is zero, some point must be so pushed into the origin.

This means that the determinant of

$$\begin{pmatrix} a & b \\ c & d \end{pmatrix}$$

is zero if, and only if, there is some point (x, y) *other than the origin* such that $(ax + by, cx + dy)$ is (0, 0), i.e. the equations

$$\begin{aligned} ax + by &= 0 \\ cx + dy &= 0 \end{aligned}$$

have a solution distinct from $x = 0, y = 0$.

This statement can be generalized for 3 × 3 matrices, and indeed for square matrices of any size. Probably the most common way in which determinants arise in algebra is by applying this condition.

AN ALGEBRAIC APPROACH TO DETERMINANTS

Starting from this algebraic condition, one can see that it is 'reasonable' that determinants should have the properties they do.

Take for example, the connexion between the determinants

$$\begin{vmatrix} a & b \\ c & d \end{vmatrix} \quad \text{and} \quad \begin{vmatrix} c & d \\ a & b \end{vmatrix}$$

If we call these D_1 and D_2 respectively, we know that

$$D_2 = -D_1;$$

this is simply Rule I. Could we have foreseen such a relation as being likely?

Let us consider the meaning of the determinants. $D_1 = 0$ is the condition for a pair of numbers, x, y, other than 0, 0, to satisfy the equations

$$ax + by = 0 \ldots (1)$$
$$cx + dy = 0 \ldots (2)$$

while $D_2 = 0$ is the condition that there should be two numbers, other than 0, 0, to satisfy the equations

$$cx + dy = 0 \ldots (3)$$
$$ax + by = 0 \ldots (4)$$

(Equations (3) and (4) are written down by looking at the letters in the second determinant.)

But the equations (3) and (4) are simply the equations (1) and (2) written in a different order. Naturally, if there is a solution x, y apart from 0, 0 when $ax + by = 0$ is written above $cx + dy = 0$, there will be a solution when $ax + by = 0$ is written below $cx + dy = 0$. That means to say, if $D_1 = 0$, it must be that $D_2 = 0$. And vice versa, if $D_2 = 0$, that is, if (3) and (4) have a solution, then (1) and (2) have a solution, and so $D_1 = 0$. Thus D_1 and D_2 must be closely related; for if either one of them is zero, the other must be. With a little care, one can prove from this consideration (for determinants of any number of rows and columns) that Rule I must be true.

There are certain snags to avoid in this proof; if one does not care to prove Rule I this way, at least one can see that Rule I is a very reasonable thing to happen.

In the same way, Rule II corresponds to the fact that it is permissible to add k times equation (2) to equation (1).

$$(a + kc)x + (b + kd)y = 0 \ldots (5)$$
$$cx \quad + \quad dy \quad = 0 \ldots (6)$$

The equations (5) and (6) say exactly the same as equations (1) and (2). It is not at all surprising that the determinants should be equal in these two cases.

Interchanging columns in the determinant would correspond to writing the y terms in front of the x terms, like this

$$by + ax = 0 \ldots (7)$$
$$dy + cx = 0 \ldots (8)$$

This changes the sign of the corresponding determinant. We should expect it only to have some simple effect on the determinant, since it is itself so trivial an alteration.

In the same way, all the properties of determinants can be shown to correspond to fairly obvious properties of the corresponding systems of equations. We have succeeded in escaping from sheer calculations, with which this chapter began, and have reached the realm of ideas.

Probably the algebraic approach allows a simpler proof than the geometrical approach via matrices; we do not have to go into the question of what volume means in space of n dimensions.

QUADRATICS WITH THREE ROOTS

In algebra it is well known that a quadratic cannot have three distinct roots. If the numbers a, b and c satisfy a quadratic equation, two of these three numbers must coincide, $a = b$ or $b = c$ or $c = a$. Let us see how this known result appears in terms of determinants.

We look for a quadratic equation $px^2 + qx + r = 0$, that is to be satisfied by a, b and c. Substituting a, b and c in the equation, we find that they will satisfy it if

$$a^2p + aq + r = 0$$
$$b^2p + bq + r = 0$$
$$c^2p + cq + r = 0.$$

But these three equations have a matrix form. They state that the matrix

$$\begin{pmatrix} a^2 & a & 1 \\ b^2 & b & 1 \\ c^2 & c & 1 \end{pmatrix}$$

sends the point (p, q, r) to the origin $(0, 0, 0)$.

Now there are two possibilities. Perhaps (p, q, r) already is at the origin, and the matrix is not doing anything unusual. But in that case $p = 0$, $q = 0$ and $r = 0$, so that the quadratic equation that a, b and c satisfy is $0x^2 + 0x + 0 = 0$, which is not much of an equation. It is satisfied by every number there is.

But if (p, q, r) is a point distinct from the origin, then the matrix must be singular. So, if a, b, c are to be the roots of a genuine quadratic (one different from $0x^2 + 0x + 0 = 0$) we must have the determinant condition

$$\begin{vmatrix} a^2 & a & 1 \\ b^2 & b & 1 \\ c^2 & c & 1 \end{vmatrix} = 0$$

An exercise in many algebra texts is to show that the determinant above equals $(a-b)(a-c)(b-c)$. The fact that it factorizes in this way is not surprising; for as we saw at the beginning, a, b, c can only satisfy a genuine quadratic equation if $a = b$ or $a = c$ or $b = c$, corresponding exactly to the three factors.

DETERMINANTS AS SOURCES OF PATTERN

Determinants often help to give shape to something that without them would be shapeless. Imagine, for example, that you were asked to generalize the well-known elementary result,

$$a^2 - b^2 = (a + b)(a - b).$$

At first the expression $a^2 - b^2$ does not seem to have much pattern. It might however strike you that it can be expressed as the determinant

$$\begin{vmatrix} a & b \\ b & a \end{vmatrix}$$

Here a certain pattern is beginning to reveal itself, and you might be led to consider the determinant

$$\begin{vmatrix} a & b & c \\ b & c & a \\ c & a & b \end{vmatrix}$$

This expression when multiplied out comes to $a^3 + b^3 + c^3 - 3abc$ and has the factor $(a + b + c)$. A generalization to 4, 5 or any number of variables is evident. Students often meet the expression $a^3 + b^3 + c^3 - 3abc$ in exercises taken from algebra textbooks. They may be puzzled why this particular expression keeps cropping up. The determinant shows that it has a very definite pattern, and may be expected to have a number of simple properties, suitable for authors of textbooks in search of material for exercises.

CHAPTER TEN

Projective Geometry

Infinity is where things happen that don't.

Statement made by a schoolboy

Projective geometry is one of the most beautiful parts of elementary mathematics.

For the professional mathematician it is undoubtedly an essential part of one's education. One does not need to go very far with it; the value of a detailed study of it is doubtful, except for the specialist. But the basic patterns of projective geometry can be traced in many other branches of mathematics; they serve to guide and to unify.

For non-mathematicians, too, it is a worthwhile study. This is not on account of its technical value. It has some connexion with the theory of aerial photography, but I would not like to advocate including it in the syllabus for that reason. Rather its value is that it enlivens a course. A non-mathematician, learning mathematics for technical reasons, often has to plough through masses of routine procedures, which can be extremely dull. These tend to drug the mind. An education should also contain elements that perform the functions of a cold bath – to provide a shock and keep one awake.

Projective geometry does this very nicely. It is surprising; it does things one would not think allowable, and gets away with them. It abounds in beautiful impossibilities. In it, parallel lines meet, and there is a theorem (sufficient to make the average man doubt the sanity of mathematicians) that all circles have two points in common. These points, of course, are no ordinary points; they are imaginary, and at infinity. Still, the result is striking enough.

Moreover, the subject is an excellent example of mathematical style. In projective geometry, if something can be proved at all, it can usually be proved simply. In this respect, it is the opposite of Euclidean geometry. In Chapter 2 we met the principle (illustrated by the Wine and Water problem) that a subject becomes both general and simple by the process of *abandoning all unnecessary information.* Now this is precisely what projective geometry does. It steadfastly ignores certain types of information which cause much of the complexity of Euclidean geometry.

Again, its history is extremely interesting, and shows how little we realize the eventual consequences of anything we do. Projective geometry began as a very practical subject, in effect as the theory of perspective. If an artist wants to draw a table or a box, how should he do it? The artist might, of course, be an engineer, and in fact G. Desargues (1593–1662), an engineer and architect, was both. It is to his work that projective geometry can be traced.

The theory of perspective in itself is interesting. It assumes of course ordinary Euclidean geometry as a starting point, and applies that geometry to the theory of drawing pictures. Projective geometry thus appears first of all as a part of Euclidean geometry. It took centuries before people realized that projective geometry was in fact an independent subject; that it was infinitely simpler than Euclid; that it could be developed without any mention of Euclid, and in fact that the best way of cleaning up Euclid (which is a vast mess of unstated assumptions) was to develop projective geometry first, and get Euclidean geometry from it. Projective geometry to-day is a clear, sharp, and logical subject, which Euclid never was.

There are problems for the teacher because of this history. Should the teacher follow the historical approach? To do so has the advantage that the pupils see and can understand how the subject grew; the whole thing is seen as a natural development. But there is an objection; Euclid is illogical, projective geometry is logical. What a pity to make the logical seem to depend on the illogical! What a pity for students to learn what they will later have to unlearn! But on the other hand, if one starts from the modern viewpoint, the work is clear and logical, but the pupil has no idea what it is all about, or where it has come from.

It seems to me that one should begin by showing pupils the historical development, but warn them, from the first, that the subject reached a stage where it had to be stood on its head. Logically, for the angel on the telephone, the modern development is undoubtedly superior; the assumptions are few and simple; there is no appeal to diagrams or to physical experience of shapes and sizes. But just because it is so suitable for logicians and angels, it is insufficient by itself for most human beings. To see the clear, logical ideas gradually being disentangled from vagueness and confusion is vastly more instructive than simply starting with the logical ideas. If the limitations of the older methods are clearly explained, I do not feel that too much 'unlearning' will be necessary. A student can distinguish between a 'proof' that would have been accepted in 1640 and one that would be acceptable in 1899 (the year when Hilbert published his *Foundations of Geometry*). Whether future centuries will find it necessary to speak of 'so-called proofs' accepted by twentieth-century mathematicians I cannot guess: probably they will. But at any rate we have advanced on the seventeenth century.

It was mentioned earlier that the practical value of projective geometry rests not on its direct technical applications to photography or drawing, but on the influence which it has had on other branches of mathematics. Many examples could be given of this influence; some of these would involve long explanations. For the moment, just one example will be mentioned. Probably the branch of mathematics most widely used by engineers and scientists is that of differential equations. Anyone who has done a course on this subject will remember how disjointed it seems to be; here is an equation that you can solve by one method, here is an equation that can be solved by another – countless different types to remember, countless different methods to use. As one disgusted lecturer said, 'It is botany, not mathematics'. (Rather unfair to botany, which, after all, aims at classification, not just the mere collection of specimens.) Now there is a theory, due to Sophus Lie (1842–1899), which establishes a single principle underlying all these different types; it shows that all the equations we know how to solve have a certain property in common; it shows why they can be solved by the methods used. This theory is obviously essential to any mathematician, who wants to help practical men by showing how to solve types of differential equation that have resisted treatment until now. It is no accident that Sophus Lie was a geometer. His ideas, which proved so powerful for differential equations, had their origin in geometrical questions, closely connected with projective geometry.

THE THEORY OF PERSPECTIVE

So much, then, for the beauty and the utility of our subject. Now let us look at the subject, as it develops from the theory of drawing.

Here is Figure 50, intended to represent a cube. The picture differs in many ways from the actual cube. The height, breadth, and length of a cube are equal; but the lines in the picture which represent them are unequal. The angles of a cube are right angles;

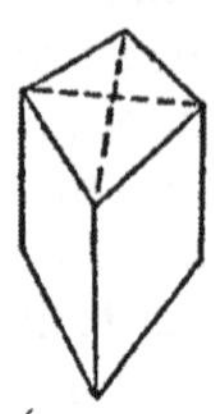

Figure 50

the corresponding angles in the picture are not right angles. Lines parallel in the cube are not parallel in the picture. Even ratios of lengths are not preserved. The dotted lines drawn on the top of the cube bisect each other in reality; they do not do so in the picture.

But yet something must be preserved. If nothing at all was preserved in going from an object to its picture, there would be no such thing as a bad drawing or a good one. And certain things are preserved. For example, if in reality a line is straight, the picture also must show it as straight. The picture of a point is a point. If a point lies on a line, the picture should also show a point lying on a line; just as, from an aerial photograph, you could tell that a certain river passed through a certain town in reality, because it appeared to do so on the photograph. (Towns admittedly are not points, nor rivers lines.)

Thus (1) being a straight line, (2) being a point, (3) being 'on', are properties that the photograph preserves. Such properties are called *projective*. A *projective property* is one that is preserved when a photograph is taken. Projective geometry is concerned only with such properties; it ignores all others.

If we take a photograph of a geometrical figure, and then take a photograph of the photograph, all projective properties of the original figure will appear in this second photograph. It does not matter how many times you go on photographing. Projective properties are preserved at each stage, and therefore are preserved throughout the whole process.

As we saw earlier, length, angle, parallelism, ratios of lengths are all altered by photography. None of these are projective properties. None of them may be mentioned in strict projective geometry.

Nearly all the results of ordinary school geometry are about lengths or angles. What is there left to talk about if these things are taboo? Are there any theorems at all?

A projective theorem, completely independent of measurement, is shown in Figure 51.

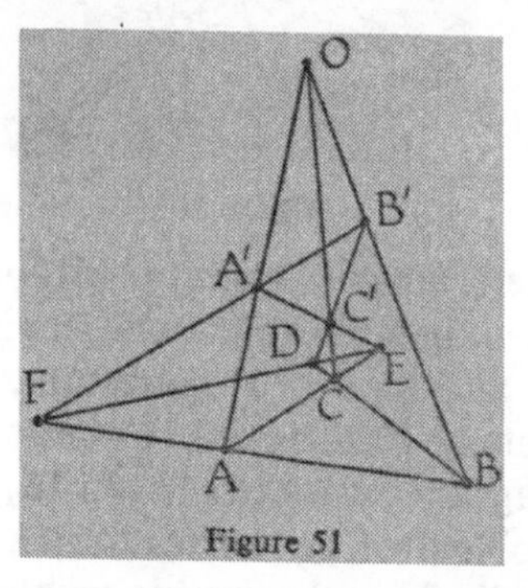

Figure 51

Starting at O, draw the three lines OA, OB, OC. A, B, C can be anywhere on these lines. Also mark any three points A', B', C'; A' on OA, B' on OB, C' on OC. Join AB and $A'B'$. These two lines meet in F. In the same way, AC meets $A'C'$ in E, BC meets $B'C'$ in D. You will now find that D, E, F lie on a straight line.

This result makes no appeal to the ideas of length or angle. It uses only the ideas of straight line and point. It is fully projective. If you took a photograph of this diagram, the photograph would do just as well as the original diagram.

This example shows that there are such things as projective theorems.

This particular result is known as Desargues' Theorem – the same Desargues as was mentioned earlier. Incidentally this theorem possesses a remarkable symmetry. Every point in it is as good as any other point. The construction I gave earlier obscures this fact. We began with O, then brought in A, B, C and A', B',

C', and finally arrived at D, E, F. But if you rub out the letters on the diagram, you will find that you can mark any one of the ten points as being O. You can then find ways of marking in the other nine letters so that the printed statement of the theorem given earlier remains exactly true. The diagram is the same, but the points are differently named. One way of re-lettering is shown here. (Figure 52.) There are in fact 120 different ways of

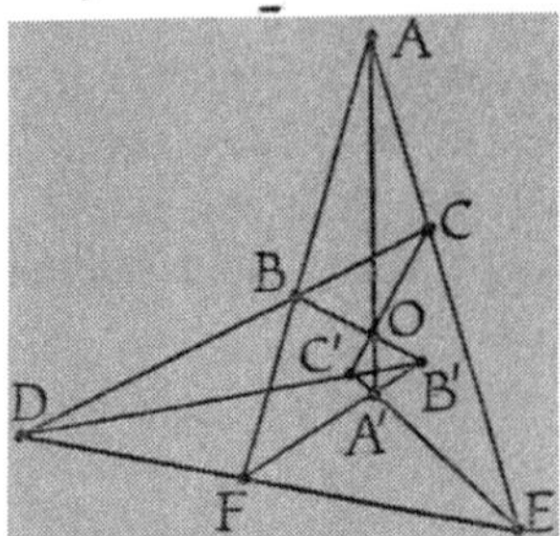

Figure 52

putting in the letters on this diagram, without any alteration in the printed statement being necessary. There is perfect democracy among the points and lines of the diagram. There are 10 lines, each having 3 points on it; there are 10 points, each having 3 lines through it. The diagram is spoken of as 'a 10_3, 10_3 configuration'.

An obvious problem of generalization; what other diagrams have similar properties? In other words, what other configurations exist? In a configuration, each line must have the same number of points on it, and each point the same number of lines through it.

Desargues' Theorem incidentally gives the answer to a nursery puzzle; plant ten trees in ten rows of three.

ORIGIN OF DESARGUES' THEOREM

If you have any knowledge of Euclid, you will see that this theorem is quite unlike most theorems of school geometry. It would make a very nasty problem, if you were suddenly required to prove it by means of Euclid (though it can be done). A more interesting question is, how did Desargues come to think of this result? Once you know how Desargues came to it, you can see that the result is obvious.

Desargues, as we saw earlier, was interested in how to make drawings of buildings and other solid objects. If you can look at

Desargues' diagram, and see it as the picture of a certain solid object, you will have the result at once.

It is not easy to imagine solid objects, or to visualize them from drawings. The best thing is actually to make them. A cheap and convenient method is to get some old newspaper, and roll each sheet up until it forms a rod or tube, about as thick as a pencil but considerably longer. Instead of rolled newspapers you can use pea-sticks if you have some to hand.

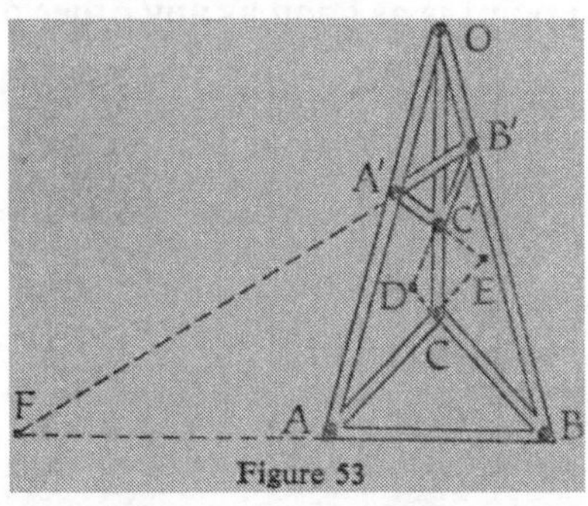

Figure 53

Take three rods, and place them to form a tripod. (Figure 53.) O is at the top. A, B, C are on the ground. A', B' and C' are next chosen, placed on the three legs of the tripod as shown. These points should not all be the same height above the ground. Rather they should be so placed that if someone were to slash through the tripod with a very sharp sword. cutting it at A', B' and C', then the sword would strike the earth not very far away from A, B and C in its follow through. Now place a rod so as to pass through A' and B', and lay another rod on the ground, touching the feet A and B of the tripod. These two rods will meet at F. In the same way rods AC and $A'C'$ will meet in a point E, and rods $B'C'$ and BC will meet at D.

The points D, E, F so found all lie on the ground. But they also lie in the plane of the slashing sword mentioned earlier. These two planes will meet in a straight line, as you can see from your model. If you now take a photograph, or make an accurate drawing, of your model, you will obtain the figure for Desargues' Theorem.

PAPPUS' THEOREM

Another theorem independent of angle and length is Pappus' Theorem. It involves the nine points and nine lines shown in

Figure 54. A natural way to draw this diagram would be first of all to draw the lines *ABC* and *DEF*, and mark the points *A*, *B*, *C* anywhere on the upper line, *D*, *E*, *F* anywhere on the lower line. Where *FB* meets *EC* gives *L*; *M* is found from *FA* and *DC*; *N* from *EA* and *DB*. The points *L*, *M*, *N* will then be found to lie on a line.

This diagram is also a configuration, being a 9_3, 9_3. There are 108 different ways of putting in the letters on this diagram, without any change in the statement of the theorem becoming necessary. Each point is as good as any other point.

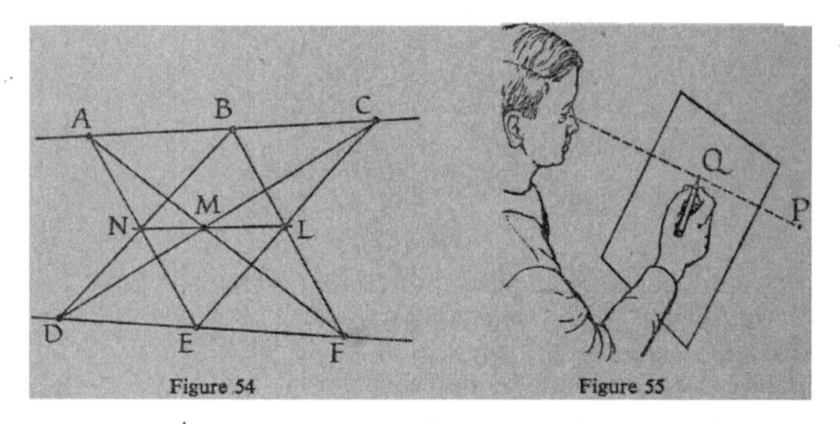

Figure 54 Figure 55

WHAT IS A PICTURE?

So far we have talked about 'taking photographs' and 'drawing pictures' but we have not explained exactly what this means. Imagine you have a sheet of glass. You wish to draw on this glass a picture of various objects behind it. You rest your head in such a way that the position of your eye is fixed. If *P* is any point of the object to be drawn, you make a mark on the glass at *Q*. (Figure 55.) This mark just hides the original point *P* from your sight; your eye, the mark *Q*, and the point *P* are in line. If you make similar marks on the glass for other points of the objects to be drawn, you obtain a picture of these objects.

Figure 56 shows the process in the reverse direction; lines drawn on a glass slide throw shadows on a screen – a simplified version of the cinema or magic lantern.

Both of these processes we shall speak of as *projection*. It is quite usual to speak of a cinema *projector*, so the use of the word in connexion with the second diagram is quite natural. In geometry, it makes no difference whether the picture is being enlarged or made smaller; we use the word projection also for the first diagram. Indeed we shall use it also for a situation like this one (Figure 57), which is something like what happens in a photographic enlarger. The details are unimportant. All that matters is that rays go from a point, to a point, or through a point; a picture is formed where these rays meet a plane.

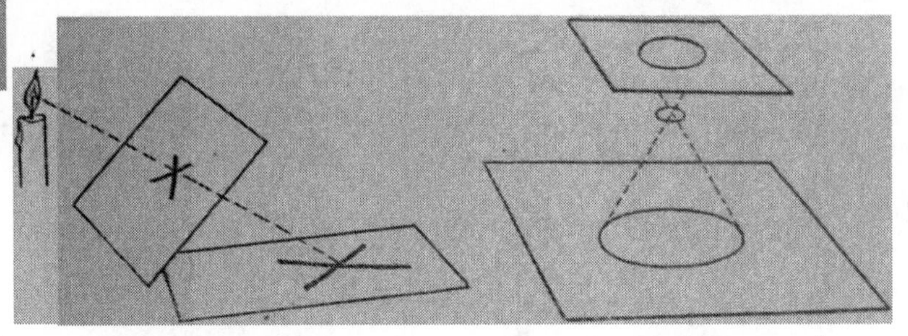

Figure 56 Figure 57

PROJECTION OF A LINE

If I mark any two points on a piece of paper, I can claim the result to be an aerial photograph showing the positions of London and Paris. If L and P (Figure 58) are the actual positions of London and Paris, and A and B are two points on my paper, I have only to place my eye at O in the figure here, and A will hide London from my eye, while B will hide Paris.

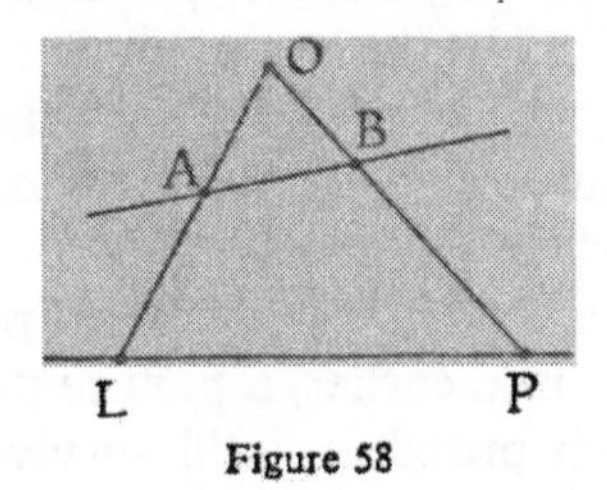

Figure 58

Two points of a line, then, have no projective property at all; no information is gained from a picture of two points – except in so far as it shows the two points to be distinct. My aerial photograph could be criticized if I said, 'B represents Paris and A represents the capital of France'.

Nor is there anything to be said about three points in a line. Suppose there are three towns, *P, Q, R* on a straight road, and I produce a line with any three points *A, B, C* marked on it. I can still claim it is an aerial photograph of the towns *P, Q, R*.

Suppose I place my picture so that *A* coincides with *P*. (Figure 59.) Now placing my eye at *O, A, B* and *C* represent the correct positions for a picture of *P, Q* and *R*. Again, I am supposing *P, Q, R* to be distinct points, and *A, B, C* also to be three distinct points.

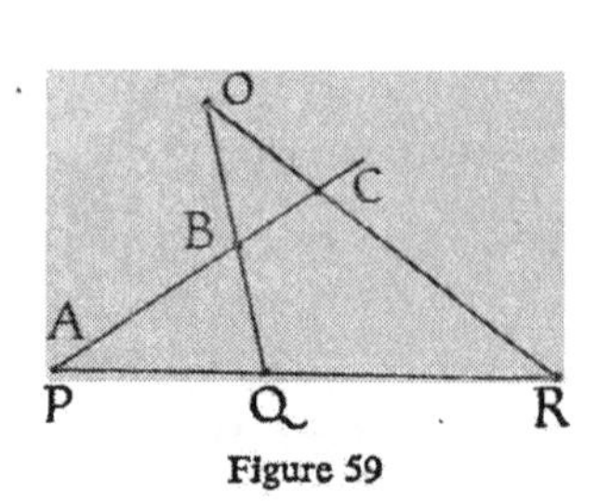

Figure 59

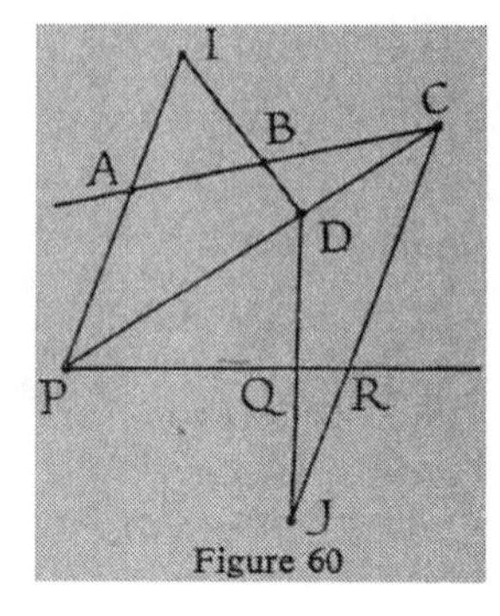

Figure 60

Any professional mathematician reading the last paragraph will be critical. The argument depends on 'placing the picture' in a certain way. What do I mean by that? I must explain, logically, what is implied in moving a picture about. (A thing, by the way, Euclid never did, and one of the reasons why we criticize him these days.) I can avoid this, however, by saying that my picture is only a photograph of a photograph of the three towns. Suppose *A, B, C* (Figure 60) is my alleged photograph. *P, Q, R* are the towns. We join *PC*, and take any point *D* on it. Let *PA* meet *DB* in *I*. Looking at things from *I*, *ABC* is a true picture of *PDC*. But *PDC* is a true picture of *PQR*. We have only to look at things from the point *J*, where *DQ* meets *CR*.

So *ABC* is a picture of *PDC*, which is a picture of *PQR*.

But as we emphasized earlier, a picture preserves all projective properties, and such properties still survive in the picture of a picture. So we see that there is nothing to be said in projective geometry about three points of a line. If somewhere in the world there were a great flat desert, on which three rocks were known to lie in line; and if an aeroplane flew over the desert and took a photograph in which these three rocks and nothing else appeared

– then no new information would be contained in that photograph; if the position of the aeroplane when it took the photograph were unknown, It would be impossible to work out anything at all about the distances between the rocks.[1]

Projective geometry has nothing whatever to say about three separate points in a line; except that there are three of them, and that they are separate.

CROSS-RATIO

But the situation is different when there are four objects in a straight line. An aerial photograph does tell you something about these.

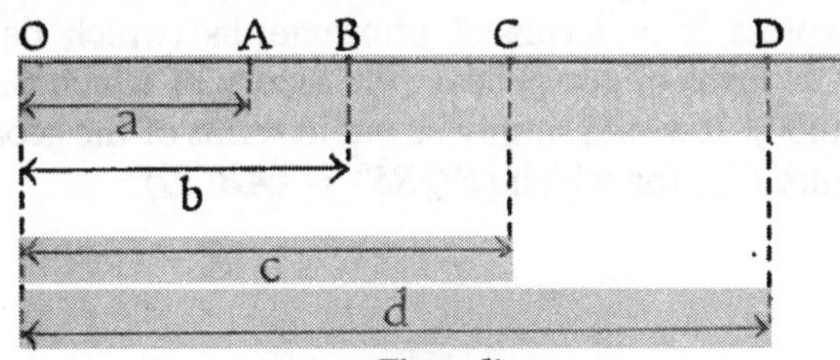

Figure 61

Suppose A, B, C, D (Figure 61) are four points in a straight line. Take any point, O, as origin and suppose that A, B, C, D are distances a, b, c, d respectively to the right of O.

We can calculate the quantity

$$x = \frac{(a-b)(c-d)}{(b-c)(d-a)}.$$

This quantity has the property that *it is exactly the same for the picture as for the original object.* If you wanted to calculate x, it would not matter if you measured the distances on the photograph, or on the actual site.

The camera can lie. It lies when it tells us that equal lengths are unequal, and that right-angles are not right-angles. The only thing it does not lie about is this expression

$$(a-b)(c-d)/(b-c)(d-a).$$

1. One could perhaps say that rock B was *between* rocks A and C. But this is due to the nature of a camera; with projection defined in the most general way, for the purposes of pure geometry, even the order of three points on a line is not preserved by projection.

The value of this quantity we can find directly from a photograph. And anything that can be asserted definitely on the evidence of a photograph can be put in terms of such quantities.

Naturally, a name is given to this quantity. It is called the *cross-ratio* of the points A, B, C, D. The sign $(ABCD)$ is commonly used as an abbreviation for this quantity.

Note that $(ABCD)$ stands for a single number. For instance, if in the diagram above $a = 2$, $b = 3$, $c = 5$, $d = 7$ then $(ABCD)$ means $(2-3)(5-7)/(3-5)(7-2) = (-1)(-2)/(-2)(5) = -\frac{1}{5}$. You notice that this number can be negative. It can also be positive, as you can verify by taking, say, $a = 2$, $b = 3$, $c = 5$, $d = 4$.

It is a remarkable fact that this expression $(ABCD)$ was already known to Pappus, who lived before A.D. 300. Naturally he did not express it in terms of photographs (which date from 1829), nor in terms of perspective (the theory of which developed after A.D. 1400). It would simply be put in terms of the geometrical figure (Figure 62), for which $(PQRS) = (ABCD)$.

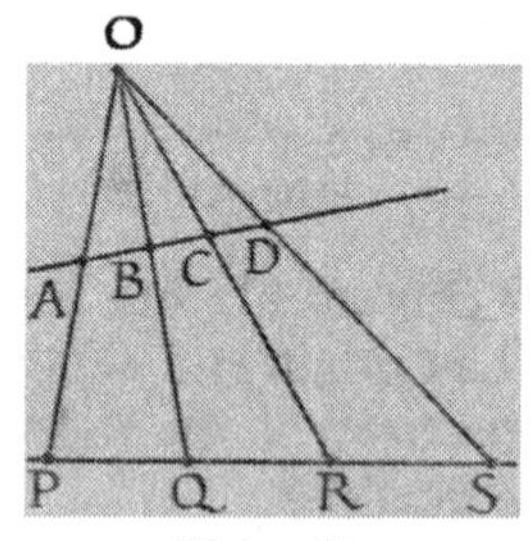

Figure 62

It is not known whether Pappus himself discovered this result, or whether he learnt it from some earlier writer. Nor do we know what train of thought led to this theorem. This is a pity; it would be extremely interesting to know how it was discovered. For it is by no means an obvious result. Imagine that we did not know about this expression, and we wanted to find some property that was the same for the picture as for the object. How would we begin to search for such a property? A modern mathematician, with the help of invariant theory, would know what to do. But for someone living in A.D. 300 to discover it – that is remarkable.

An example of how the cross-ratio can be applied to drawing; I have tried here to draw (Figure 63) four lamp-posts, evenly spaced along a road. Have I drawn them correctly?

If in the picture I measure from the foot of the nearest lamp-post, I find $p = 0$, $q = 6$, $r = 8$, $s = 9$. So the cross-ratio $(PQRS)$ is $(0-6)(8-9)/(6-8)(9-0)$, that is, $-\frac{1}{3}$. For the four points A, B, C, D which are equally spaced, as the lamp-posts should be in reality, we have $a = 0$, $b = 1$, $c = 2$, $d = 3$. The cross-ratio is $(0-1)(2-3)/(1-2)(3-0)$, which is again $-\frac{1}{3}$. As the cross-ratio is the same in the two cases, it follows that P, Q, R, S can be regarded as a correctly drawn picture of A, B, C, D.

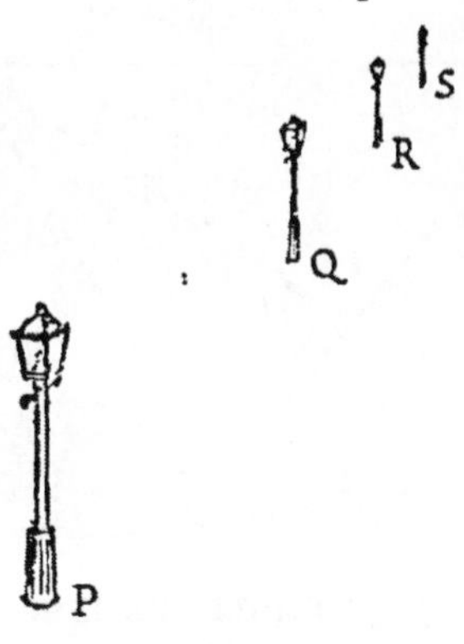

A B C D

Figure 63

This calculation, you will realize, does not check how well I have drawn the lamp-posts themselves, but only whether I have spaced out the points P, Q, R, S, representing the feet of the lamp-posts, correctly.

THE HORIZON

One of the first things an art student learns is how to draw parallel lines. Figure 64 shows how parallel lines appear in a picture. The picture represents, say, a lawn with a path all round it. The shape *ABCD* is intended to represent a rectangle as it appears to the eye. *EFGH* is also intended to represent a rectangle. *AB, EF, GH, DG* are intended to represent parallel lines. But the art student will not draw them as parallel. They will be drawn so that if they were produced (as shown by the dotted lines) they would meet at the point *Q*. In the same way, *AD*, *EH*, *FG*, *BC* are supposed to represent parallel lines; they meet at a point *P*, in the picture. The points *P* and *Q* lie on the horizon. They are rather peculiar points. Any other point in the picture represents an actual point. *E* represents a corner of the lawn; *D* represents a corner of the path. But *P* and *Q* do not represent actual points. It would be no good to say, 'We will have a picnic to-day at the point represented by *Q*'.

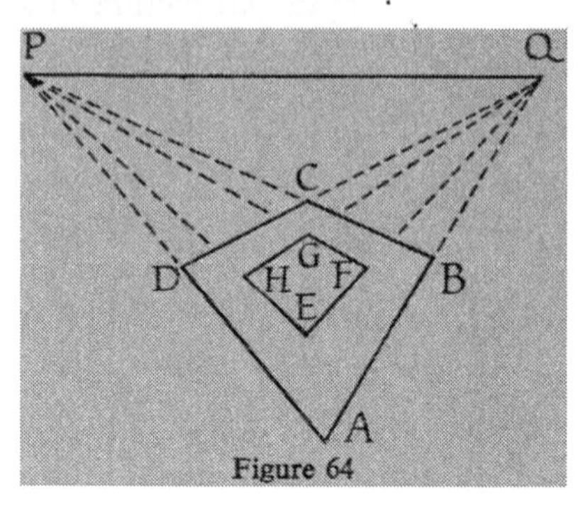

Figure 64

The most we can say *Q* represents is a *direction*. Any line, drawn on the ground parallel to *AB* will appear to go through *Q*. In the same way, *P* represents the direction *AD*. Any line on the ground parallel to *AD* will appear to go through *P*. Any other point on the horizon also represents a direction.

In the picture we therefore have two kinds of points, (i) honest to goodness points, which really represent places, (ii) points on the horizon, where parallel lines *seem* to meet.

Lines which all meet at a point are known as concurrent. Thus, in a picture, lines which in reality are parallel, appear concurrent. This suggests that parallel lines and concurrent lines must have many properties in common; in fact, *all projective properties*, for these are the properties common to the picture and the reality. Accordingly, mathematicians have come to regard parallel lines as if they were a special case of concurrent lines ('mathematics is the art of giving the same name to different things'). In fact, we talk as if *P* and *Q* really did represent something. We have got into the way of saying 'parallel lines meet at infinity'; *P* is spoken of as representing 'a point at infinity'. All such points together form 'the line at infinity', represented by the horizon *PQ* on the picture.

The advantage of this manner of speaking is that it enables us to unify theorems. Instead of saying, 'these two lines either meet or are parallel' we can say 'these two lines must meet'; but of course, it may be that they meet 'at infinity' – which is a new way of saying that they do not meet!

In Desargues' Theorem or Pappus' Theorem it may well happen that some of the lines, which earlier I assumed to intersect, are in fact parallel. By speaking of them as meeting at infinity, I can still interpret these theorems and obtain a result that is true even when the lines are parallel. You may care to draw these figures, with some of the lines parallel, and see what happens to the theorems.

You may have doubts about whether it is justified to speak of points at infinity as if such things actually existed. It is right that you should have such doubts, and thinking about such questions helps to form one's philosophy. Later I will say something about the logic of the subject, that may, or may not, satisfy you. For the moment, I suppose we are merely exploring. We try this way of speaking to see where it leads. If we like the results, we will adopt the habit of speaking so.

RATIOS OF LENGTHS IN PICTURES

Suppose we look at a picture and on it we see three points A, B, C which are in line, and which represent stones lying on the ground. Can we, from the picture, deduce anything about how these stones are actually situated on the ground?

We certainly cannot deduce anything from the three points A, B, C because, as we saw earlier, in projective geometry any three points on a line are as good as any other three distinct points. They have no projective properties. But suppose the horizon is shown on the picture, meeting the line AC in D, so that D represents the point at infinity on the line of the stones. We now have four points A, B, C, D on the line, and they do tell us something. The cross-ratio $(ABCD)$ in the picture will be the same as the cross-ratio of the four points on the ground. Let a, b, c measure the positions of the stones along the line on the ground. These distances will be measured from some fixed point of that line, it does not matter which; D represents a point at an infinite distance along the line, so we must take $d = \infty$. No respectable analyst would dream of actually substituting an infinite value for a symbol, still less of doing what we shall do in a minute. However what we do here can be justified *for projective geometry*.[1] We substitute the value $d = \infty$ in the expression $(a-b)(c-d)/(b-c)(d-a)$ $c-d$ then becomes $-\infty$, while $d-a$ is $+\infty$. Cancelling (!) these we find the value $-(a-b)/(b-c)$ for the cross-ratio. This is the same thing as $-(b-a)/(c-b)$. Now if P, Q, R (Figure 65) represent the actual positions of the stones on the ground, $b-a$ measures the distance PQ, while $c-b$ measures the distance QR. Accordingly

$$-(b-a)/(c-b) = -PQ/QR.$$

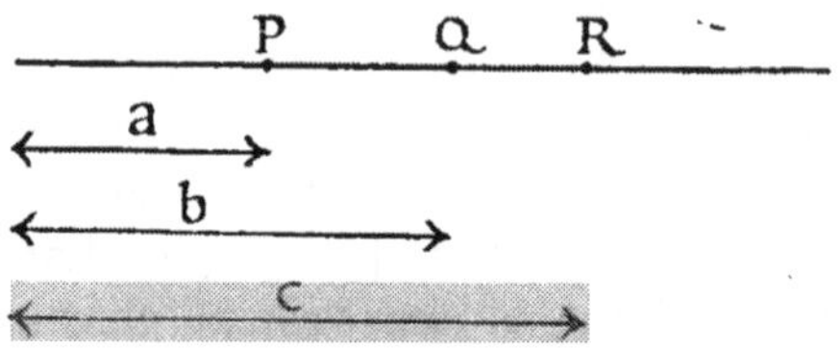

Figure 65

That is to say, the cross-ratio of P, Q, R and the point at infinity on the line on the ground is equal to $-PQ/QR$: that is to say, it is the ratio in which Q divides PR, with a minus sign. But the cross-ratio is faithfully preserved in the picture. Accordingly, if in the picture we measure the lengths AB, AC, AD and deduce the cross-ratio $(ABCD)$, it will have a minus sign, and its size will tell us the ratio in which the middle stone divides the line joining the other two stones.

In particular, if Q is the mid-point of PR, then $PQ = QR$, $PQ/QR = 1$, and $(ABCD) = -1$. Four points, A, B, C, D, on a straight line for which $(ABCD) = -1$ are said to form a *harmonic range*.[2] Harmonic ranges thus arise naturally in connexion with the theory of perspective. A harmonic range gives a picture of a bisected line, together with the point at infinity on the line. Harmonic ranges play a considerable role in the develop-

1. In projective geometry there is a line at infinity; in inversive geometry a point at infinity; in analysis infinity is taboo. In fact the word 'infinity' means three different things in these three subjects, and it is a pity that we have the same word for such different uses.

2. The name 'harmonic' is due to the fact that the lengths AB, AC, AD then have a certain connexion with the theory of musical instruments.

ment of projective geometry in its modern sense (that is, as a subject depending on neither the theory of perspective nor the idea of length).

CONSTRUCTION OF HARMONIC RANGES

Earlier we considered how to draw a picture of a lawn. Suppose (Figure 66) *ABCD* is our lawn, in the shape of a rectangle. *AC* and *BD* cross at *R*. *RS* is parallel to *AD*. Clearly *S* is the mid-point of *AB*.

Now suppose we draw a picture of the lawn. It will look like this.

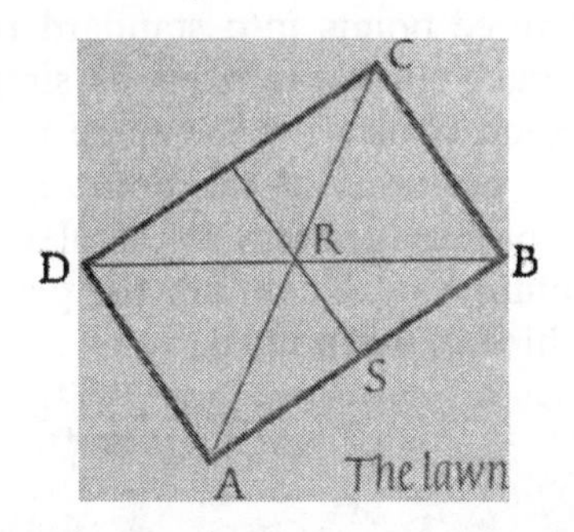

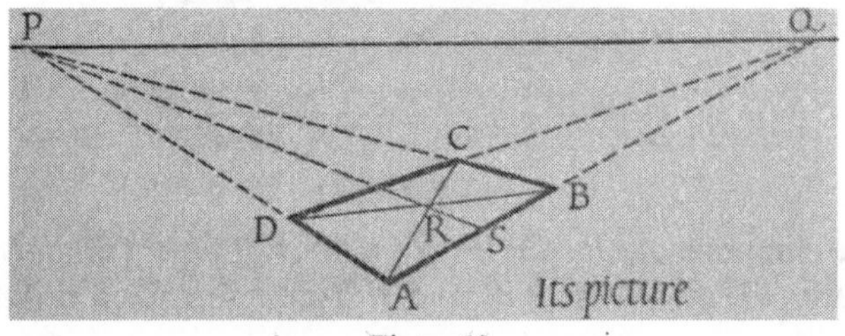

Figure 66

The points *P* and *Q* of course are on the horizon. Since *ASBQ* is a picture of a bisected line together with a point at infinity, *ASBQ* must be a harmonic range, i.e. $(ASBQ) = -1$.

Now we can forget about the original lawn, and look at this last diagram simply as a geometrical figure, not as a picture of anything. It gives us a construction for a harmonic range. We only need a ruler to draw it: it consists purely of straight lines; no measurement of lengths or angles comes into it. It is a purely projective construction for a harmonic range. If we took a photograph of this diagram, the photograph would do just as well as the diagram.

GEOMETRY APPLIED TO ALGEBRA

In Chapter 3 cross-ratio was mentioned, and it was suggested that the reader should do some rather heavy algebra to verify its properties. Now we come back to the same question, but armed with the knowledge that cross-ratio is something unaltered by projection. We also know that three points can be projected into any three points we like. By means of this we can simplify the work considerably.

We shall project three points into standard positions, which we will choose to make the expressions as simple as possible. What would be a good choice for the three points, or for the numbers that fix the positions of the points? 0 is an obvious choice. So is 1. Both of these are easy to calculate with. The third point we select is infinity, ∞. These are the places to which we shall try to project three of our points.

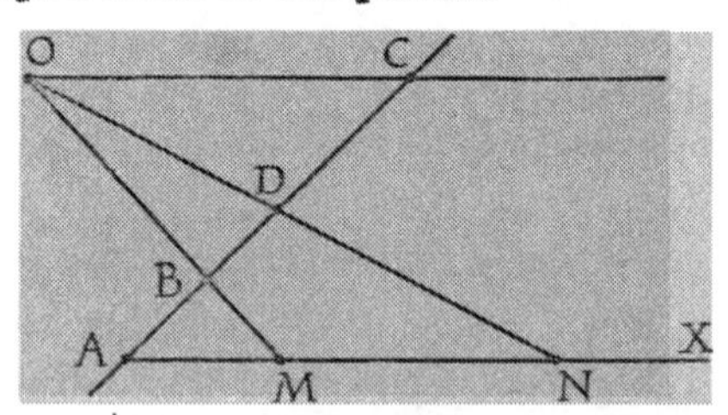

Figure 67

We suppose A, B, C, D to be four points taken on a line (Figure 67), at distances a, b, c, d from some fixed point of that line.

Through A draw another line, AX. Let a point N be chosen on AX so that the distance AN equals 1. Join N to D. Through C draw a line parallel to AX, meeting ND in O. O is the point we shall project from, so that the shadows of the points, A, B, C, D, fall on to the line AX. As OC is parallel to AX, C is projected to infinity, as desired. A is already on AX and projects into itself. D projects into N. If we measure from A, A is at distance 0, N is at distance 1. If we use the letter Q to stand for the projection of C, Q being at infinity has the symbol ∞ attached to it. Finally the projection of B is at M. We use x to stand for the distance AM.

Since cross-ratio is unaltered by projection,

$$(ABCD) = (AMQN),$$

or, in terms of numbers $f(a, b, c, d) = f(0, x, \infty, 1)$. But $f(0, x, \infty, 1)$ is easily worked out; with the treatment of ∞ that we used earlier, we have $f(0, x, \infty, 1) = (-x)(\infty)/(-\infty)(1) = x$.

In the same way, if we alter the order of the letters, we have, for example, $(ABCD) = (AMNQ)$, so

$$\begin{aligned} f(a, b, d, c) &= f(0, x, 1, \infty) \\ &= (-x)(-\infty)/(x-1)(\infty) = x/(x-1). \end{aligned}$$

This is one of the set that we met in Chapter 3, but this time it comes straight out; there is no heavy algebra, no thinking what is the best way of verifying the result.

This is a striking example of the benefit we can derive from knowing that a function is unchanged by a certain procedure; the procedure in this case being that of projection.

In studying any problem it is therefore wise to ask, 'What procedures leave this problem unaltered?' The greater the variety of procedures we can find that leave it unaltered, the greater freedom we shall have in manœuvring it into a manageable form. In this last example, our freedom consisted in the fact that we could manœuvre three of the symbols occurring in $f(a, b, c, d)$ to the three positions chosen by us in advance, 0, 1 and ∞.

There is a moral here of very wide application.

A function unaltered by a particular procedure is called *invariant* (= unvarying). Thus $f(a, b, c, d)$ is *invariant under projective transformation.*

But projection is still a geometrical idea. If we can translate this geometrical notion into algebra, we shall have a purely algebraic result, something that we can verify by algebra without any appeal to geometrical constructions.

In Figure 68 we have an example of projection. The lines OX and HJ are fixed; so is the point E. F moves along the line HJ. Its shadow, thrown on to OX by a light at E, gives the point G.

If we denote the distance HF by the letter t, and the distance OG by u, then the quantity u has been obtained from the quantity t by means of a geometrical construction involving projection, i.e. by a projective transformation. Our present aim is to see what sort of algebraic function gives u in terms of t.

But to find u in terms of t is a straightforward piece of co-ordinate geometry. We simply have to run through the geometrical construction, and at each stage translate what is happening into algebra.

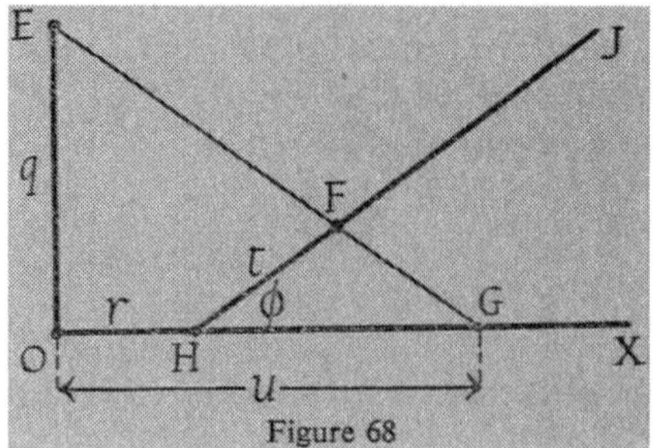

Figure 68

We suppose O to be taken as origin, E to be the point $(0, q)$, H the point $(r, 0)$, and the line HJ to make an angle φ with OX. All this information is marked in the diagram. The quantities q, r, φ of course are constants. The only things that vary are t and u. You can, if you like, think of t as standing for *time*, so that the point F would be moving along HJ according to the law $s = t$, that is, with unit velocity. The point G moves also, being the shadow of F, and we want to find the law of its motion.

At any time t the point F has the co-ordinates

$$(r + t\cos\varphi,\ t\sin\varphi),$$

as is easily seen on dropping a perpendicular from F to HG.

Now we know the co-ordinates of E and of F. There is only one line joining these two points, and it is purely routine to find its equation. There are several different ways of finding the equation. Various methods or formulae are available. As we considered this problem in Chapter 4, we may as well quote equation (III) of that chapter, in which we must put $a = r + t\cos\varphi$, $p = t\sin\varphi$ these being the co-ordinates of F, and $b = 0$, for the x co-ordinate of E. The letter q already has the correct meaning, being the y co-ordinate of E. We thus find the line EF to have the equation

$$y = x\frac{t\sin\varphi - q}{r + t\cos\varphi} + q.$$

G is where this line meets OX, which has the equation $y = 0$. On putting $y = 0$ in the equation above, and solving, we find the x co-ordinate of G must be

$$u = \frac{q(r + t\cos\varphi)}{q - t\sin\varphi}.$$

This result looks much more complicated than it really is. The only thing that is varying is t. We are only interested in how t appears in this equation. (This is the same point that was made in Chapter 4.) If we carefully choose numerical values for the constants q, r, φ, the simplicity of the formula will appear. We may, for example, choose for φ the angle 53° 8′ which has a simple sine and cosine; in fact, $\cos\varphi = \frac{3}{5}$, $\sin\varphi = \frac{4}{5}$. Taking $q = 1$, $r = 2$ we have as an example of a projective transformation

$$u = \frac{10 + 3t}{5 - 4t}.$$

With other values for q, r, φ the numbers would not be quite so simple arithmetically, but they will always be constants, and thus always we have a relation of the form

$$u = \frac{g + ht}{k + lt} \ldots \text{(P T.)}$$

The example selected above would correspond to $g = 10$, $h = 3$, $k = 5$, $l = -4$.

The letters P.T. above stand for 'Projective Transformation'. We see that a very simple algebraic process corresponds to the geometrical idea of projection.

Geometry thus calls our attention to the type of algebraic function labelled (P.T.) above. This function has important properties and arises in many parts of mathematics which have no obvious connexion with geometry.

For example, imagine we were studying the integral

$$\int \frac{dx}{\sqrt{(x-a)(x-b)(x-c)(x-d)}}$$

Integrals can very often be dealt with by writing $x = F(z)$; if the function $F(z)$ is carefully chosen, the integral may come to a simpler form. This is the well-known device of change of variable, in elementary calculus.[1] But what function $F(z)$ are we to choose? One might well be at a loss for an idea.

But if we look at the integral, we notice that the four numbers a, b, c, d come into it. Evidently, something special happens for

1. See, for example, Fawdry and Durell, *Calculus for Schools*, Chapter 14.

$x = a$, $x = b$, $x = c$, $x = d$. We saw earlier that any three points could be projected into any three points. The corresponding algebraic result is that we can always find a projective transformation that sends any three numbers to any three numbers. The three numbers usually chosen (as we saw earlier) are 0, ∞, 1. This means that we can choose the constants g, h, k, l in the equation

$$x = \frac{g + hz}{k + lz}$$

in such a way that $x = a$ when $z = 0$, $x = b$ when z equals (or tends to) infinity, $x = c$ when $z = 1$.

There is no difficulty in finding the constants, but I need not give the result here. We suppose the constants to be found, and then substituted in the equation above. We then use this equation to replace the variable x by the variable z in the integral.

The details of this work can be carried out by anyone with a knowledge of elementary algebra and calculus, and the willingness to cover a couple of pages of paper with symbols. No difficulty of principle arises.

For anyone who does not wish to carry out this work, the interest of these remarks will lie in the general idea involved; that the solution of a problem in calculus may be helped by the use of an algebraic function suggested by the geometrical theory of aerial photography. This strikingly illustrates the interdependence of the various branches of mathematics; the fact that there is a subject *mathematics*, which is not merely a collection of technological applications.

The result of the detailed calculation is to reduce the integral to the form

$$C \int \frac{dz}{\sqrt{z(z-1)(z-f)}}$$

The expression under the square root is now only of the 3rd degree, instead of the 4th. (This is due to the fact that a point has wandered off to infinity.) The quantity f that occurs above is what we earlier denoted by $f(a, d, b, c)$, the cross-ratio of the numbers a, b, c, d. Again, geometry helps us: we have met $f(a, d, b, c)$ in geometry. We therefore *recognize* this function when it appears in a calculus problem. If we had not first of all studied projective geometry and seen the significance of cross-ratio, we should have thought nothing of it if the expression $(a-d)(b-c)/(d-b)(c-a)$ occurred in the course of a calculation. It would just have been another collection of symbols. But now we say, 'Ah, the cross-ratio! Why does that occur here?'

TRANSFORMATIONS

The example just considered shows the usefulness of transformations, for the classification of problems. The two integrals that appear in the last section look quite different. One contains a quartic, the other a cubic. But projective transformation changes the first into the second, and, if we want to reverse the process, it is quite easy to change the second back again into the first. Each integral therefore can be regarded as *the other in disguise*. Anything we know about one of them tells us a corresponding fact about the other. In fact, we soon come to regard them, not as two different problems, but as two different forms of the same problem.

This is the great value of transformations. They cause problems which, at first sight, we should regard as distinct, to merge into a single problem. Having solved a problem, we not merely know the solution of that problem, but also of *all the problems which can be considered as that problem in disguise.* Some transformations are fairly feeble in their effect. They may provide only *one* disguise for each problem. In that case, they double our knowledge. To every problem we have solved, the transformation provides a mate. Problems are thus classified in pairs. But some transformations are much more powerful than this. The transformation may contain one or more constants, which we can choose at will. We are then able to transform our problem into an infinity of different shapes. Knowing its solution, we know the solution of a whole family of other problems.

Here again is a growing point for mathematics. We have found one transformation, the projective transformation, which has a certain interest and value. At first we may be content simply to use this transformation, which enables us to classify problems. But later we begin to think, 'This particular type of transformation has helped us. It has multiplied our knowledge. What other kinds of transformations are there?'

We seek in fact not merely to generalize our *results*; we desire also to generalize our *methods*. Having found a certain method useful, we look for other methods with similar properties. Having found a particular transformation useful, we look for other transformations, we ask what their properties will be, we try to classify them. In view of the power that transformations have, the study of transformations is obviously a valuable line of enquiry.

The unending nature of the mathematical process should be evident. We first classify problems. That is easy enough to grasp. We then classify ways of classifying problems. That sounds more complicated. There is a third step in the process, but it sounds too much like a tongue twister to be worth printing. And beyond that are fourth and fifth steps, each one classifying what has been achieved previously.

The process has no termination. But each stage of the process yields a certain satisfaction; it enables us, with a few principles, to survey a wide field of knowledge. We shall return to the study of transformations in another chapter.

This chapter began with a reference to all circles having two common points. This we have not yet discussed, nor have we dealt with the logical justification of the line at infinity. Both points will be dealt with in the next chapter.

CHAPTER ELEVEN

On Apparent Impossibilities

Reason has moons, but moons not hers
Lie mirrored in the sea,
Confounding her astronomers
But, oh, delighting me.

Ralph Hodgson

You have only to show that a thing is impossible and some mathematician will go and do it.

A saying

In the last chapter we used some highly questionable arguments about infinity, and left the way open for all kinds of philosophical disputes, such as, 'Is there *really* a place, infinity, where railway lines meet? If not, is it justified for mathematicians to speak of infinity, and of parallel lines as meeting there?'

The answer to these two questions is, or very well may be, (1) No, (2) Yes. Of the correctness of the second answer I am certain; the mathematical use of 'the line at infinity' is justified. The first question belongs not to mathematics but to physics; it is a question about what happens at the edges of the universe. I do not know if the universe has any edges; if it has, I have not been there, and I do not know what they are like, nor how railway lines, if transported thither, would behave.

This answer may leave you with the feeling that somehow you are being cheated. We are brought up to think of geometry as telling us the truth about things. What is to be said of a geometry that claims to be true, even if things behave quite otherwise? Even for a mathematician too it is a little disturbing. We do not want it to be thought that mathematics has *nothing* to do with truth.

Mathematics in the first place is concerned with consistency of argument. A mathematician can say to a physicist, 'Here is a consistent theory. Whether it actually fits the physical universe, I do not know. But I do believe it to be consistent with itself; it will answer *yes* or *no* to any question you put to it, and it will never answer both *yes* and *no* to the same question'.

The job of the physicist is to test whether such a mathematical theory fits the observed facts; whether the pattern embodied in the mathematics coincides with a pattern occurring in actual life.

It is in fact found that this procedure is extremely fruitful; that mathematical arguments do in fact lead to practical results. I am not sure that this could be predicted by mathematics alone. The universe might be chaotic. That it is orderly, that it can be mastered by means of logical thinking, this seems to me to be a result of experience, incapable of logical or mathematical proof.

When we say a mathematical procedure is justified, we mean only that this procedure, however much you use it, will not lead you to contradict yourself. We assert its *formal consistency*; nothing more. Yet even this, I suppose, is a statement about reality; we say that however much you use this argument, you will never be led to contradict yourself, provided of course that you use it properly. In the course of centuries this statement might be found false; we should then abandon the argument in question, and no longer recognize it as a valid mathematical procedure.

In this chapter, then, I have to try to show that a consistent scheme can be produced, in which parallel lines meet at infinity. The consistency of this scheme is accepted by all living mathematicians.[1] Whether future ages will find any loophole in it is for you to judge.

HOMOGENEOUS CO-ORDINATES

We shall find it easier to discuss points at infinity if we use a particular device, known as homogeneous co-ordinates. This device is very helpful throughout projective geometry, and also in other branches of mathematics. It may not strike you as a very natural procedure; your first impression may be that it is making things more complicated, because instead of specifying a point of a plane, as usual, by two numbers (x, y), we shall be using three numbers (X, Y, Z). There are however many places in plane geometry where the symmetry of the algebraic expressions arising suggests the desirability of having three basic numbers instead of

1. In practice at any rate. Among mathematical philosophers the consistency of arithmetic is still a matter for discussion. The consistency of arithmetic being granted, that of projective geometry follows.

two: I cannot give examples of this without going into greater detail than is here desirable.

The new numbers, X, Y, Z are connected with the old co-ordinates x, y by the equations

$$x = X/Z, \; y = Y/Z,$$

which are certainly simple enough.

For example, if we want to express the point $x = \frac{7}{8}$, $y = \frac{3}{8}$ in the new system, we can take $X = 7$, $Y = 3$, $Z = 8$. As you can see from the equations above, this will give us the desired values for x, y. I say here, 'we can' and not 'we must' choose the values 7, 3, 8 for X, Y, Z. For, since $\frac{7}{8}$ is the same as $\frac{14}{16}$ and $\frac{3}{8}$ the same as $\frac{6}{16}$, we could just as well have taken the values 14, 6, 16, or in fact any other three numbers having the form $7k$, $3k$, $8k$.

This is the first peculiarity of homogeneous co-ordinates. You can multiply the quantities X, Y, Z by any number you like without altering the position of the point they represent.

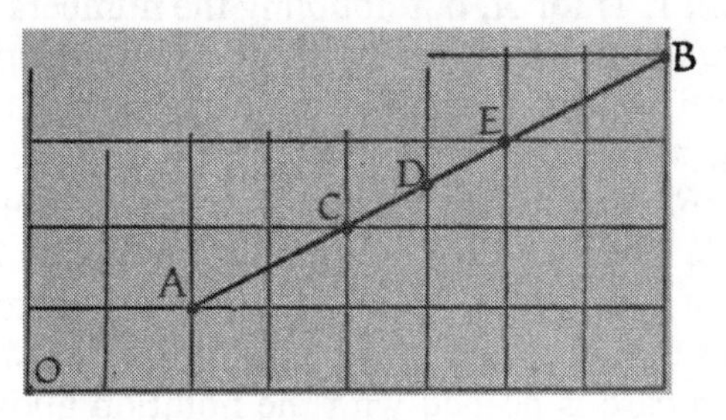

Figure 69

We are now in a position to see something which works out very simply in the new co-ordinates. In Figure 69, the point A has $x = 2$, $y = 1$, while B has $x = 8$, $y = 4$. In the new system we could take A to be (2, 1, 1) and B to be (8, 4, 1). Suppose now we were to add these numbers together; will this give us anything of interest? By adding together, I mean that we write the numbers belonging to B below those belonging to A, and do three addition sums.

A is	(2, 1, 1)
B is	(8, 4, 1)
Addition gives	(10, 5, 2)

This process is easy to perform; has it any geometrical meaning? If we go back to our familiar x, y system we see that (10, 5, 2) corresponds to $x = 5$, $y = 2\frac{1}{2}$. These are the co-ordinates of D, the mid-point of AB.

I always think of D as being a *mixture* of A and B – something like blending two kinds of tea in equal quantities. Let us see what happens if we vary the proportions. Suppose we take two of A to one of B.

The numbers for A, doubled, are	(4, 2, 2)
The numbers for B, as they stand	(8, 4, 1)
Addition gives	(12, 6, 3)

this result corresponds to $x = 4$, $y = 2$, the point C which is one-third of the way from A to B. The point obtained is still on the straight line, but doubling the contribution from A has pulled the point towards A. You can easily verify that taking the original numbers (2, 1, 1) for A, but doubling the numbers for B, leads on addition to the point E, two-thirds of the way from A to B.

A natural way of recording these results with the minimum of writing would be as follows: $D = A + B$, $C = 2A + B$, $E = A + 2B$. As you can verify for yourself, $A + \frac{1}{2}B$ also gives the point C. The blend of tea obtained by mixing 2 lb. of Ceylon with 1 lb. of China tea (if such a mixture is ever made) is exactly the same as that obtained by mixing 1 of Ceylon with $\frac{1}{2}$ of China.

A certain care is needed with the notation above. Do not fall into the following fallacy. $2A + 0 \cdot B$ leads to the point A. So A and $2A$ both represent the same point. Thus $2A = A$. Add B to both sides. $2A + B = A + B$, that is, $C = D$. But C and D are distinct points. The fallacy of this argument is clearly seen in terms of tea blending. A cup of tea made from a 2 lb. packet of Ceylon tea tastes exactly like a cup made from a 1 lb. packet. It does not follow that a blend of 2 lb. of Ceylon with 1 lb. of China will taste like a blend of 1 lb. of Ceylon and 1 lb. of China.

With these equations, one must only use arguments that hold in regard to proportions of mixtures. The proportion 1 : 0 and the proportion 2 : 0 give the same type of mixture (the first ingredient alone); but 1 : 1 and 2 : 1 give different mixtures.

Coming back to our diagram on the graph paper, we saw that the points $A + B$, $2A + B$, $A + 2B$ all lay on the line AB. By choosing suitable positive numbers m and n, we can get $mA + nB$ to lie anywhere we like on the line between A and B. (You may find it instructive to work out a number of examples. Where, for instance, is $99A + B$? Where is $49A + 51B$?) To get outside the stretch AB we need to use negative numbers. For instance, if we take $-4A + B$, we make the following calculation.

-4 times the numbers for A,	$(-8,$	$-4,$	$-4)$
the numbers for B,	$(\;8,$	$4,$	$1)$
Addition gives	$(\;0,$	$0,$	$-3)$. So $x = 0, y = 0$.

$-4A + B$ thus represents the origin, O, which as you can see from the diagram, lies on the line AB.

If you work out in turn $-3A + B$, $-2A + B$, $-1\frac{1}{2}A + B$, $-1\frac{1}{4}A + B$, $-1\frac{1}{8}A + B$, you will find that you get points lying on the line BA, if it is extended to pass beyond O. Each point lies further to the left than the previous one. The number that goes with A is getting nearer and nearer to -1. If we actually take the value -1, and consider $-A + B$, we get the set of numbers (6, 3, 0) for X, Y, Z. We are now unable to find x, y. Our standard equations $x = X/Z$, $y = Y/Z$ lead to the meaningless expressions $6/0$, $3/0$.

If we pass the value -1, no further difficulty arises; it is perfectly easy to work out x and y. When we took the sequence of values $-4, -3, -2, -1\frac{1}{2}$, etc., approaching -1, the corresponding point moved down the line getting further and further to the left. By coming close enough to -1, you can make this point go as far away from the origin as you like. What do you think will happen when the value -1 is passed? Where does $-\frac{7}{8}A + B$ lie? Where does $-\frac{1}{2}A + B$ lie? (As only arithmetic is needed to answer these questions, I leave them to you. It is quite interesting to follow the path of the point $kA + B$, as k passes from a large negative value, say $-1{,}000{,}000$, towards -1, through -1 and on to 0, and then up to a large positive value, say $+1{,}000{,}000$. Does it pass through all the points of the line AB? Does it end near where it began?)

THE TELEPHONE AGAIN

In all the above work, we found that our (X, Y, Z) label gave us a perfectly good (x, y) label for a point, except when $Z = 0$.

Here we come to the parting of the ways. If we decide that the

(x, y) label is the important thing, we shall have to say that three numbers (X, Y, Z) specify a point except when $Z = 0$. But suppose we have no prejudices in favour of (x, y); then we may equally well say that (X, Y, Z) *always* represents a point, and regard it as *a defect of the* (x, y) *system* that it provides no label for points with $Z = 0$.

If we were explaining this to our friend the angel over the telephone, we should find him equally prepared to accept either path. It would perhaps depend on how we put the matter. Suppose we were approaching a celestial contractor with the following specification for the creation of a universe:

I. The universe is to contain points specified by numbers (X, Y, Z).

II. Only the ratios of these numbers are to be significant. The point with the label (kX, kY, kZ) is to be identical with the point (X, Y, Z), whatever k.

III. There is to be a point with the label (X, Y, Z) whatever numbers X, Y, Z may be; except that in no case is Z to have the value 0.

IV. All other details to be at the discretion of the architect.

As you know, architects are rarely willing to accept specifications. We can imagine our angel objecting, 'I don't much like the look of Clause III. Our firm has a reputation for creating only universes of mathematical elegance. Aren't you being rather arbitrary in picking on Z? If you wanted to exclude (0, 0, 0), I would be inclined to agree with you, because 0:0:0 does not establish a proportion for a mixture, but I see nothing wrong with 6:3:0'.

If we gave way completely, the result would be a universe with projective geometry. If the angel gave way completely, the result would be our usual Euclidean geometry. If a compromise was established, by which a notice was erected on all points having $Z = 0$, 'Out of bounds. This point is at infinity', the result would be (I think) Euclidean geometry regarded as a particular case of projective geometry.

LINES AND POINTS

In the preceding chapter we saw that there were projective theorems, theorems using only the ideas of *point* and *line*, points *on* lines, lines *through* points. We must specify what we understand by a line, if such theorems are to occur in the universe we are designing.

How is a line specified in the usual (x, y) geometry? It is a well-known result in elementary co-ordinate geometry that the equation $ax + by + c = 0$ always represents a line, and that every line can be represented in this way. (The better known form $y = mx + c$ fails for vertical lines.) If we substitute $x = X/Z$, $y = Y/Z$ in the above equation, and multiply by Z, we arrive at the equation $aX + bY + cZ = 0$.

This argument is based on our usual geometry. Our new universe, of course, need not have the same laws as the old one. The above argument is not therefore meant to *prove* anything. But it does *suggest* something. If our new universe is to be helpful for the understanding of the old one, it should at least be similar to the old one. We decide to carry over the form of the equation above. We accordingly agree to the following *definition of a line*: all the points (X, Y, Z) which satisfy a given equation $aX + bY + cZ = 0$ are said to form a line. The points that satisfy the equation will be spoken of as being 'on the line'; the line will be said 'to go through these points'.

All of this will go over the telephone quite nicely, and the angel will know what we mean if we speak of points on a line. We cannot of course define a line as the shortest distance between two points, because we are designing a projective universe, in which there is no such thing as *distance*. The inhabitants of this new universe will be completely unable to attach any meaning to the word. The reason for throwing out the idea of distance is that, as we saw in Chapter 2, the *less* you have in a subject, the simpler and the more powerful that subject becomes.

Now suppose we have two lines, say

$$aX + bY + cZ = 0 \ldots \text{(I)}$$
$$pX + qY + rZ = 0 \ldots \text{(II)}$$

Will these two lines meet? In other words, is there a point that is on them both? It is easily seen that if you take $X = br - cq$, $Y = cp - ar$, $Z = aq - bp$, these values satisfy both equations (I) and (II)[1]. So there is always a point where any two lines meet. In projective geometry there is no such thing as parallel lines.

1. The angel suggested, and we accepted, that $X = 0$, $Y = 0$, $Z = 0$ should not be regarded as a point. If it were regarded as a point, it would

In Euclidean geometry how do lines manage to be parallel? Let us take two parallel lines in the usual (x, y) system, say $y = x + 1$ and $y = x + 2$. If we translate these into the (X, Y, Z) system by means of $x = X/Z$, $y = Y/Z$, they become

$$X - Y + Z = 0 \text{ and } X - Y + 2Z = 0.$$

These lines meet where $X = 1$, $Y = 1$, $Z = 0$. It is owing to the fact that $Z = 0$ that we cannot find a point (x, y) lying on both lines.

Accordingly, if we decide to disqualify all the points having $Z = 0$, and to say that two lines which meet at a point where $Z = 0$ will be regarded as not meeting, we get back from projective geometry to a geometry with parallel lines in it.

We do this in effect every time we look at a picture, and see lines which appear to meet on the horizon; we automatically disqualify the points of the horizon, they do not represent 'real' points.

Incidentally, it is worth noticing that the disqualified points are given by $Z = 0$, which is a linear equation, the equation of a line. For this reason we speak of 'the line at infinity'. Geometrically, this squares with the fact that the horizon appears straight.

From the viewpoint of projective geometry, the line at infinity is no different from any other line. If you have a piece of paper you can rule *any* straight line on it, and say, 'That is to represent the horizon'. Of course, you must view the picture in such a way that the selected line becomes level.

Projective geometry is very simple to specify. Any three numbers X, Y, Z specify a point, provided they are not all zero. An equation $aX + bY + cZ = 0$ specifies a line (provided a, b, c are not all zero). Any two points can be joined by a line; any two lines meet in a point.

This is a simple scheme. To anyone used to this scheme our ordinary geometry would appear as follows. A particular line is singled out from the rest. Although this line is exactly like all the others, a special name is given to it. It is called the line at infinity. A special name is given to two lines that happen to meet on this line. They are called parallel. Points on the special line are treated

be on every line whatever, which would be inconvenient! A difficulty would appear to arise if the solution given in the text reduced to (0, 0, 0). But it can be shown that this only happens when the lines (I) and (II) are identical, and then of course all the points of (I) lie on (II) also.

as if they did not exist. For this reason, parallel lines are said not to meet.

Euclidean geometry is thus obtained from projective geometry by disqualifying the points of a particular line. (Figure 70.) You can see that there is only one line through a given point P parallel to a given line AB. The line PQ is only called parallel to AB if PQ and AB meet in a disqualified point. All the disqualified points lie on the special line, the line at infinity (l_∞ for short). The only disqualified point on AB is C, the point where l_∞ cuts AB. So, if PQ is to be parallel to AB, to meet AB in a disqualified point, it must meet AB in C. Joining P to C thus gives the line through P parallel to AB. Clearly this construction gives one and only one line.

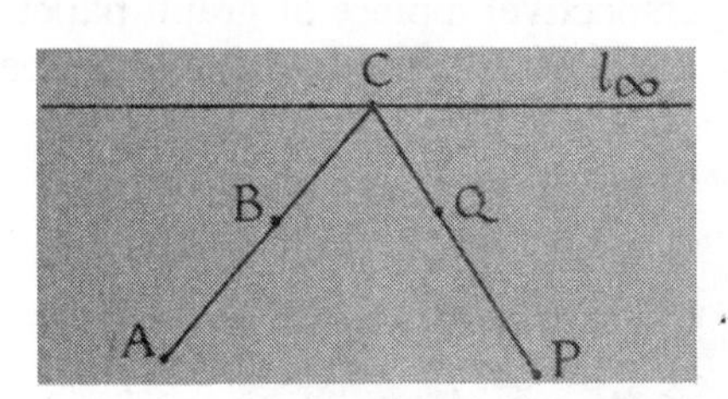

Figure 70

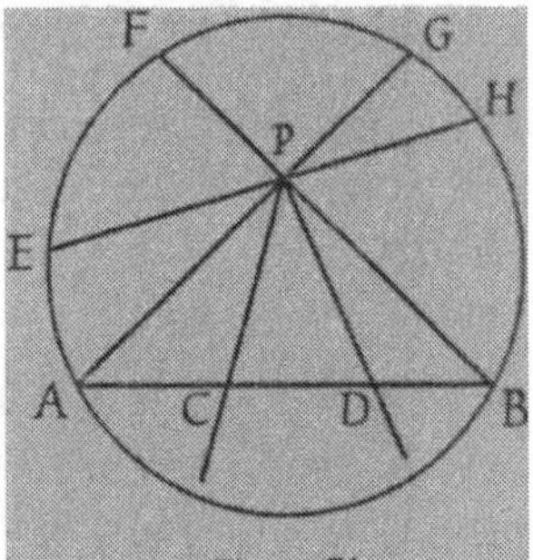

Figure 71

OTHER GEOMETRIES

In Chapter 6 we saw that the possibility of geometries other than Euclid's was not a purely theoretical speculation, but might have importance in physics. The procedure we have just followed suggests a way of getting other geometries. Why should we disqualify the points of a line? Why not choose some other curve, or region? In Figure 71 suppose that we start off with an ordinary sheet of paper, and disqualify all points lying on or outside a certain circle. This circle we can, if we like, call 'the circle at infinity'; mathematicians more commonly call it 'the absolute'. Two lines are only regarded as meeting if they meet *inside* the circle. The lines PC and PD thus count as meeting the line AB. But PA counts as parallel to AB, for it meets AB on the circle, that is, 'at infinity'. If the line PA swings round to the position PE it is still not meeting AB. If it continues to swing round in a clockwise direction, by the time it has reached *PF* it will once again be 'just' parallel, since *PF* meets *AB* at *B*. Any further clockwise rotation will cause the line to meet *AB* at a 'qualified' point. There are thus a whole bundle of lines through *P* that do not meet *AB*.

This type of behaviour will no doubt remind you of Poincaré's universe, which was discussed in Chapter 6. It is in fact Poincaré's universe very thinly disguised. By projecting Poincaré's universe on to a suitably placed sphere, and then back again on to a plane, you can obtain the geometry just discussed.

THE COLOUR TRIANGLE

The Figure 72 represents (in perspective) a piece of graph paper. *O*, as usual, is the origin. *OA*, *OB* represent the two axes, *AB* the horizon.

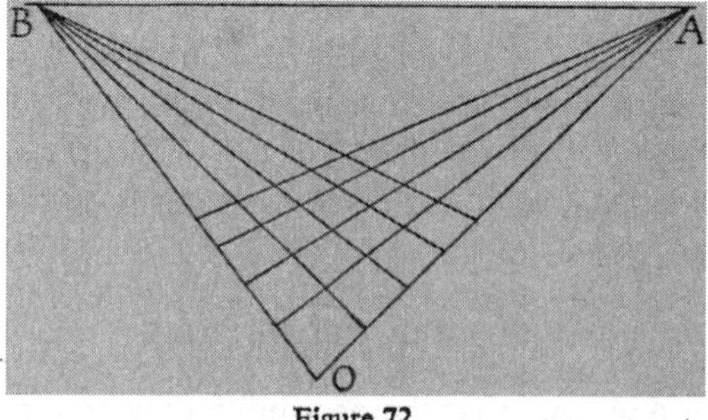

Figure 72

In the usual co-ordinates x, y *OA* would be the line $y = 0$, *OB* the line $x = 0$. In homogeneous co-ordinates, the equations are practically the same. Since $y = Y/Z$, *OA* has the equation $Y = 0$. Similarly, *OB* is $X = 0$.

The horizon, *AB*, represents the line at infinity, which has the equation $Z = 0$. The figure thus shows a *threefold* symmetry. Ordinarily, we think of graph paper as having a twofold symmetry. There are two axes, the x-axis and the y-axis, each as good as the other. But in our picture we have three lines with equal claims, *OB* labelled $X = 0$, *OA* labelled $Y = 0$, *AB* labelled $Z = 0$.

Thus the triangle *OAB* appears as the basic figure. *O* is the point (0, 0, 1), *A* the point (1, 0, 0), *B* the point (0, 1, 0). You can check the agreement of this with our previous ideas, by substituting these values in $x = X/Z$, $y = Y/Z$. O comes without any difficulty, as $x = 0$, $y = 0$. For A we find $x = \infty$, $y = 0$, so that A should be an infinite distance along the x-axis, which it is. Similarly B is $x = 0$, $y = \infty$, an infinite distance along the y-axis, as required.

In the usual co-ordinates, a line through the origin is $y = mx$. In homogeneous co-ordinates we have, for a line through O, $Y = mX$. What about lines through A? These represent lines parallel to the x-axis, i.e. lines of the type $y = c$. In homogeneous co-ordinates this equation becomes $Y = cZ$. Similarly, lines through B have equations of the form $X = kZ$. Again, there is symmetry; lines through O, lines through A, lines through B are all treated fairly; the same type of equation is given to each.

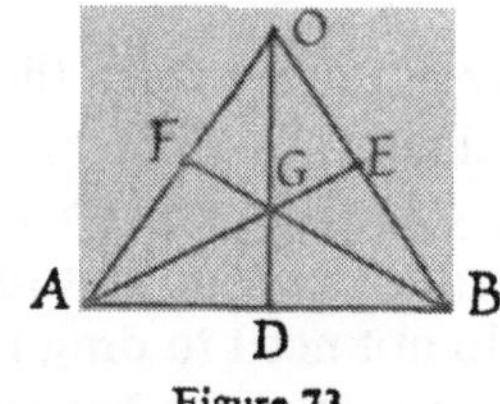

Figure 73

Since A has co-ordinates (1, 0, 0), B has co-ordinates (0, 1, 0), and O has co-ordinates (0, 0, 1) we can regard any point (X, Y, Z) as being a *mixture* of A, B and O. We have only to take X times the co-ordinates of A, Y times those of B, Z times those of O, and add. 'Mixture' is here used in the same sense as earlier in the chapter, where the 'mixing' of points was compared to the blending of tea.

This same idea of mixing is used in the Colour Triangle. The three-colour printing process depends on the fact that any colour can be obtained by mixing, in suitable proportions, the three primary colours red, blue, and yellow. The effect of this mixing can be illustrated by the colour triangle. (Figure 73.) Suppose we put pure red at A, pure blue at B, pure yellow at O. At D, midway between A and B, we put the tint obtained by mixing a pint of red with a pint of blue paint. Similarly, E shows the colour got by mixing yellow and blue in equal amounts, F that for red and yellow in equal amounts. G, in the centre, shows the effect of mixing 1 of red with 1 of blue and 1 of yellow.

Calling G the point (1, 1, 1), D the point (1, 1, 0) and so forth, the co-ordinates X, Y, Z of any point give the proportions in which the three basic colours are to be mixed at that position. This again shows that only the ratios of X, Y, Z are significant. The point (2, 2, 2) for instance is the place where 2 pints of red are mixed with 2 of blue and 2 of yellow: but this is exactly the same tint as is obtained from 1 pint of each. (2, 2, 2) is the same point as (1, 1, 1).

This type of triangular diagram is also used by chemists, when they wish to show the effect of mixing three ingredients in various proportions.

Anyone familiar with mechanics will see yet a third way of visualizing the diagram. The point (X, Y, Z) is the centre of gravity of X pounds placed at A, Y pounds placed at B, and Z pounds placed at O.

We have completely got away from the way in which we first introduced the co-ordinates X, Y, Z, by means of graph paper. We can now start directly from *any* triangle as a basic triangle. This gives us much greater freedom. If we are attacking a problem about a triangle, we do not need to drag two lines at right angles into the question. We can take the triangle itself as the basis of co-ordinates.

We are still far from having reached the most general type of homogeneous co-ordinates. The idea of centre of gravity, too, is a *metrical* one; it rests on the idea of length. It is possible to develop homogeneous co-ordinates in a purely projective manner, taking any four points of a plane as (1, 0, 0), (0, 1, 0), (0, 0, 1) and (1, 1, 1), and simply by drawing straight lines to spread a network over the whole plane, each point of the network receiving a label (X, Y, Z) without any appeal being made to the idea of length at all. Details of this procedure will be found in books on projective geometry, under the heading of Möbius nets.

IMAGINARY POINTS

In our specification for a universe, the first clause read 'I. The universe is to contain points specified by numbers (X, Y, Z)'. Nothing was said about the nature of these numbers. Up till now we have, without saying so, been treating these numbers as *real numbers*. But there is no special reason why they should be real. All our geometrical results are obtained by algebraic calculations.

Provided the numbers used for X, Y, Z satisfy the laws of algebra, we can go ahead. Now it is well known that *complex numbers*, numbers of the form $a + ib$ where $i = \sqrt{-1}$, do satisfy all the laws of algebra. So we could have a universe in which every point had a label (X, Y, Z), the numbers X, Y, Z being complex numbers. This universe would be no harder to handle mathematically than when X, Y, Z were supposed to be real, since complex numbers behave exactly like real ones in algebraic work.

But of course this universe would be very different from the geometry we are accustomed to. With complex numbers, every equation has a root. The geometrical translation of this is – any two curves meet! But in ordinary geometry it is easy to draw, for example, two circles that do not meet.

We can however overcome this divergence by adding an extra clause; Ia. The creatures living in the universe shall only be able to perceive points for which X, Y, Z are real.

So now we have two kinds of disqualification for points. (i) the points with $Z = 0$ are called at infinity, and are not capable of being reached by the creatures, (ii) points with complex values for X, Y, Z cannot be perceived at all.[1] They are called imaginary points.

The two types of disqualification are different. Two lines that meet at infinity are called parallel; but two curves that meet in imaginary points are not called parallel. The creatures simply say, 'These curves do not meet'.

What advantage is there in this curiously elaborate procedure of bringing in points with imaginary co-ordinates, only to shut them out again?

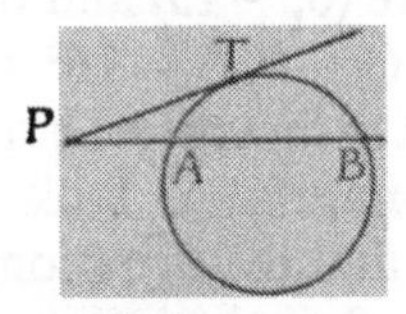

Figure 74

Its usefulness may be shown by the following example. A well-known theorem of Euclid is that, for Figure 74, $PT^2 = PA.PB$.

1. Strictly speaking, it is only the ratios $X{:}Y{:}Z$ that need to be real. The point (i, i, i) is the same point as (1, 1, 1) and would count as a point the creatures could perceive. But as it is the same point as (1, 1, 1) we should have no occasion to write it with complex values.

Now suppose we have two circles, each passing through the points A and B. (Figure 75.) From P, any point on AB, tangents PS and PT are drawn to the two circles. The theorem above gives

$$PS^2 = PA.PB = PT^2,$$

so that $PS = PT$. This is a well-known result in school geometry; the tangents from P to the two circles are equal, for any point P on the line AB.

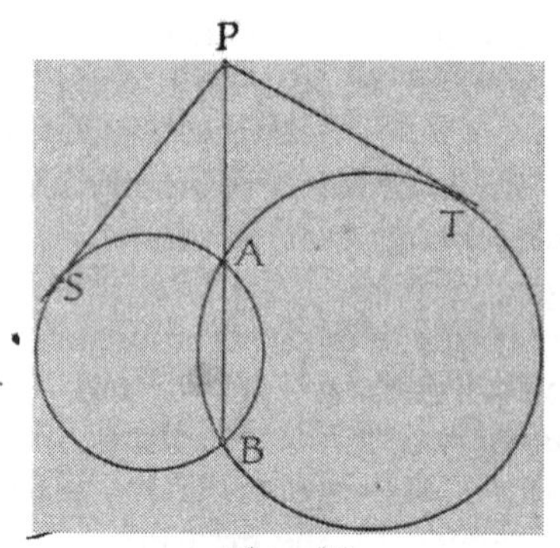

Figure 75

The proof is simple, but the annoying thing – at the level of school geometry – is that it only works when the circles meet. The result – that there is a line, such that the tangents from any point P of it to the two circles are equal – is still true when the circles do not meet. But our method of proving it has evaporated.

Now let us look at it algebraically. We will take two circles that do not meet, say the circle with centre (0, 20) and radius 16, and the circle with centre (0, −15) and radius 9. The equations of these circles are $x^2 + (y-20)^2 = 16^2$ and $x^2 + (y+15)^2 = 9^2$. These two circles do not in fact meet; but suppose we thought they did. It would then be natural to look for the points A and B, where they met, and to find the equation of the line AB, on which P would have to lie. To find the points A and B we should solve the equations as simultaneous equations. This would lead us to the solutions $x = 12i$, $y = 0$ and $x = -12i$, $y = 0$. You can easily check that these values satisfy both equations, i.e. the points $(12i, 0)$ and $(-12i, 0)$ lie on both circles. These, then, must be A and B. Both of them lie on the line $y = 0$.

So the line AB seems to be $y = 0$.

Now $y = 0$ is a perfectly good, real line. Moreover, if you draw the two circles on graph paper, and take any point P on the line $y = 0$, you will find that the tangents from P to the two circles are in fact equal. If you are sufficiently familiar with co-ordinate geometry, you will be able to prove this result by calculation, and not merely verify it.

What we have done amounts to this: we have carried out the calculations we would have made if the two circles met in real points; we have gone ahead even though $\sqrt{-1}$ turned up in our calculations; and we have ended with a correct result, in which $\sqrt{-1}$ did not appear.

That is to say, we no longer bother whether the circles meet in real points or not; we carry out the same calculations in either case. This principle, applied throughout geometry, saves an enormous amount of messing around with special cases. In a problem with three circles, for instance, all the circles might meet; or none of them might meet; or two might meet each other, but not the third; or one might meet both the others, but the others not meet. By allowing imaginary meeting points, we avoid all this detail.

Do not get worried by philosophical speculations about whether these imaginary points 'really exist'. Mathematics deals with patterns, not with things. If we can show that the pattern of the universe which incorporates Clause Ia is the same as the pattern of Euclid's plane, that is all we need do. Whether such a universe actually exists anywhere does not affect the logic of the method. It is sufficient if, without logical contradiction, it *could* exist.

THE CIRCULAR POINTS AT INFINITY

In the previous section, for the problem of the two circles, we used the ordinary x, y co-ordinates. As the line at infinity did not come into the question, that was sufficient. The only novelty was that complex numbers came in.

We are now going to ask, 'Where does a circle cut the line at infinity?' Since infinity comes in, we shall have to use X, Y, Z. Since circles do not actually go off to infinity at all, the points in question must be imaginary; complex numbers must come into the answer.

In ordinary x, y co-ordinates, the circle with centre (a, b) and radius r has the equation $(x-a)^2 + (y-b)^2 = r^2$. In homogeneous co-ordinates, introduced by means of $x = X/Z$, $y = Y/Z$ this equation becomes $(X-aZ)^2 + (Y-bZ)^2 = r^2Z^2$.

The line at infinity is $Z = 0$. Where the circle meets the line is found by putting $Z = 0$ in the equation of the circle. This gives $X^2 + Y^2 = 0$. So $Y^2 = -X^2$. Taking the square root, $Y = iX$ or $-iX$. We can take any value we like for X, since it is only the ratios $X:Y:Z$ that matter. We may as well take $X = 1$. We thus find the points $(1, i, 0)$ and $(1, -i, 0)$ to be the intersections. These points are usually referred to as I and J.

The remarkable thing about the answers $(1, i, 0)$ and $(1, -i, 0)$ is that these do not depend at all on the values a, b, r which distinguish one circle from another. That is to say, it does not matter where the centre of the circle is, or what the radius of the circle is, the circle always meets $Z = 0$ in the same two points, I and J. This is a very unexpected result. It clears the way for an entirely novel definition of a circle.

A straight line always has an equation of the form

$$aX + bY + cZ = 0.$$

Each term contains *one* of the quantities X, Y, Z. The expression is called of the first degree. The natural thing to study after lines would be curves of the second degree; the equation is now to contain *two* quantities X, Y, Z in each term. Any such curve is called a *conic*. Its equation will be of the form

$$aX^2 + bY^2 + cZ^2 + 2fYZ + 2gZX + 2hXY = 0.$$

It is not hard to show that if a conic passes through the points I, J it must be a circle.

We are thus led to the quite new definition of a circle: a circle is a conic passing through the two points I and J. This definition, though it may seem a strange one, is in fact clear cut and much simpler to handle mathematically than the vague ideas of Euclid's geometry. It is the starting point for the modern development of the geometry of circles.

CHAPTER TWELVE

On Transformations

> The art of reasoning consists in getting hold of the subject at the right end, of seizing on the few general ideas that illuminate the whole, and of persistently organizing all subsidiary facts round them. Nobody can be a good reasoner unless by constant practice he has realised the importance of getting hold of the big ideas and hanging on to them like grim death.
>
> *A. N. Whitehead*
> Presidential Address to the London Branch of the Mathematical Association, 1914

In Chapter 10 we discussed the idea that one problem might be another problem in disguise. The process of putting on, or taking off, a disguise is known as *transformation.* Obviously transformations have the effect of multiplying our knowledge, and are useful for economizing effort.

Let us first consider one or two very elementary examples of transformation. These will not yield any surprising results, but will simply show the meaning of transformation in its simplest and barest form, stripped of all complications. Later we will go to the other extreme, in order to show the great and unexpected power transformations can have.

Let us consider then the two following questions, (i) find the square root of 2, (ii) find the square root of 200.

Question (i) is simply a matter of looking up a table of square roots, in which we find that $\sqrt{2}$ is approximately 1·4142. If we look in the table for the square root of 200 we do not find it; as a rule, only numbers between 1 and 100 are listed in tables of square roots. The reason of course is that $\sqrt{200}$ is just 10 times $\sqrt{2}$. Multiplying by 10 is such a simple operation, that it would be most wasteful actually to print extra tables for numbers larger than 100. In fact, the principle just used to find $\sqrt{200}$ can be used to find the square root of any number, however large or small. An infinite extension of the tables is thus provided by that principle. Along with $\sqrt{2}$ we have all the disguised forms $\sqrt{200}$, $\sqrt{20{,}000}$, ... $\sqrt{0{\cdot}02}$, $\sqrt{0{\cdot}0002}$, ... unlimited in number.

The above argument for finding $\sqrt{200}$ could be expressed in the following algebraic form. To solve the equation (ii) $x^2 = 200$ we apply the *transformation* $x = 10y$ which reduces the problem to that of solving (i) $y^2 = 2$.

To be useful, a transformation should be simple to apply. Here, in our example, we only have to multiply by 10, which is easily done. Some transformations, naturally, are more complicated than this, but generally speaking we are quite satisfied if the transformation is simple *in comparison with the problem being transformed.* In a problem on integration, we should regard as simple any transformation that could be carried out by purely algebraic processes – for instance, the projective transformation applied to an integral towards the end of Chapter 10.

TRANSFORMATIONS AND EQUATIONS

It is easy to generalize the process just considered. Given any equation, is it possible to find a transformation that will bring it to a simpler form?

At school we learn to solve quadratics by a variety of methods – by factors, completing the square, or by formula. All of these methods we are *told.* Let us look at the question of solving a quadratic as if we were the first people ever to encounter that problem.

First of all, consider what would be involved in making tables for the solution of quadratics, supposing we decided simply to tabulate all the solutions. It is easy to make a table containing a thousand entries – think of an ordinary table of logarithms, which directly gives the logarithms of all numbers from 1·00 to 9·99, nine hundred entries usually filling a couple of pages. But if we wish to tabulate something depending on two quantities, each of which takes a thousand values, there will be a million entries. For example, if we have two variables a and b, each running from 0·00 to 9·99, we would need to have a first sheet on which a was 0·00 and b had the values from 0·00 to 9·99; then a second sheet on which a was 0·01, and b ran through the values from 0·00 to 9·99; and so on, up to a thousandth sheet on which a was 9·99 and b ran from 0·00 to 9·99.

Thus a table involving two variables would need not one sheet but a book of a thousand leaves.

In the same way a table involving three variables would require a thousand books, each with a thousand leaves. All of this is well known to makers of tables.

Now a quadratic equation contains three constants a, b, c, being $ax^2 + bx + c = 0$. Tables for the solution of quadratics would at first sight seem to require the formidable library mentioned above. Some relief is given by the fact that we can divide by a, and thus make the coefficient of x^2 unity. This leaves us with an equation of the form $x^2 + px + q = 0$. Even so, we still have two quantities p, q and a volume of a thousand pages appears necessary.

Can we improve the situation by means of transformations? That is, can we make the equation simpler by putting x equal to some expression containing a new variable y?

Take for example the equation $x^2 - 3x - 5 = 0$. Our transformation, as was mentioned earlier, should be a simple one. As the problem itself is quadratic, this suggests that the transformation should be something even simpler, that is, linear. We might try the general linear expression, that is, $x = my + c$. However it turns out to be sufficient simply to take $x = y + c$. On substituting this in our equation we have, after multiplying out and collecting terms $y^2 + (2c-3)y + c^2 - 3c - 5 = 0$. The simplest thing that occurs here is the coefficient of y, namely $2c-3$. If we choose c to be $1\frac{1}{2}$, this coefficient will become zero.

Accordingly we find that the equation $x^2 - 3x - 5 = 0$ is simplified by the transformation $x = y + 1\frac{1}{2}$, which in fact reduces it to the equation $y^2 - 7\frac{1}{4} = 0$.

This last equation is the same as $y^2 = 7{\cdot}25$ and can be solved straight away by consulting a table of square roots. The fact that the fraction $\frac{1}{4}$ occurs in the reduced equation, while there were no fractions in the original equation, is immaterial. It is as easy to look up the square root of 7·25 as the square root of 7. We can then deduce the solutions of the original equation by adding on $1\frac{1}{2}$.

In fact any quadratic equation can be reduced by a suitable transformation to a form in which it is only necessary to consult a table of square roots.

Thus, so far from requiring the fantastic collection of volumes which at first seemed necessary, we require only our usual table of square roots in order to solve any quadratic equation whatever.

This particular example illustrates the role which transformations play and the economy which they effect. The example is elementary, but the principle involved applies equally well in many advanced branches of mathematics.

A SIMPLE TRANSFORMATION OF A GRAPH

A natural thing to consider in connexion with any shape is the possibility of moving that shape to another position. For instance, it is well known that the graph of $y = x^2$ is a U-shaped curve like that shown in the figure. In Figure 76, A is at the origin, P is any point of the curve, Q is immediately below P, and the relation $PQ = AQ^2$ holds, since PQ is y and AQ is x.

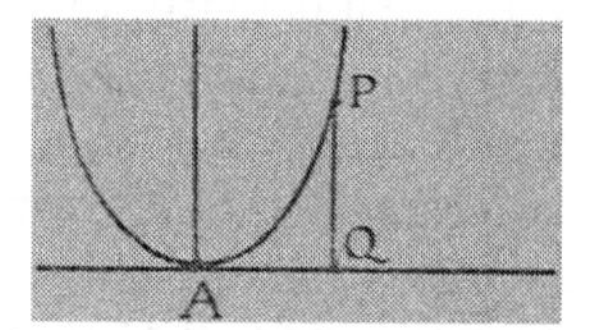

Figure 76

Suppose we trace this figure with tracing paper, and then move the tracing paper until the point A is over the point (3, 2) on the graph paper. The situation is then as shown in Figure 77. What will be the equation of the curve in its new position?

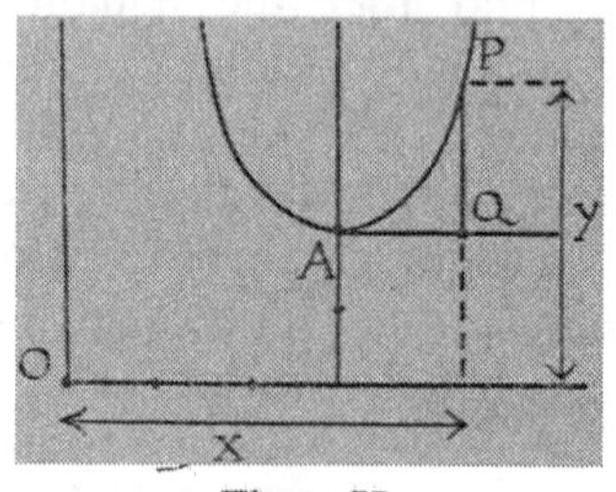

Figure 77

Since the curve is still of the same shape, we still have $PQ = AQ^2$, but it is no longer correct to say that AQ equals x. The distance P is to the East of the origin O is not AQ now, but $3 + AQ$, and $AQ = x-3$. In the same way, $PQ = y-2$.

Accordingly, the property $PQ = AQ^2$ translated into algebra now becomes $y-2 = (x-3)^2$, which on simplification gives

$$y = x^2 - 6x + 11.$$

That is to say, if you have drawn the graph of $y = x^2$ and you are asked to draw the graph of $y = x^2 - 6x + 11$, you do not need to make any calculations. The new graph is simply the old one moved 3 units to the East and 2 units to the North.

It is not difficult to show that any graph with an equation of the form $y = x^2 + px + q$ is simply the graph $y = x^2$ displaced a certain distance from its original position.

If you know all about the graph $y = x^2$, you know all about all the graphs of the form $y = x^2 + px + q$.

CONFORMAL TRANSFORMATIONS

In Chapter 1, under the heading 'Nature's Favourite Pattern', a list was given of a dozen or so subjects of practical importance, which had a certain connexion with $\sqrt{-1}$.

Certain problems arising in these subjects have the property that they can be transformed into a whole host of other problems, so that if we know the solution of one problem, we can immediately deduce the solutions of a whole family of other problems.

The transformations used are known as conformal transformations. First of all I will try to explain what such a transformation consists in, and then an example will be given showing the great power of this method.

To every function $f(z)$ – or at any rate, to every 'reasonable' function – there corresponds a conformal transformation. The function might be $z^3 - 7z + 2$, or e^z, or $\log z$, or $\sqrt{z+3}$, or many another, it does not matter. To each of these functions there corresponds a transformation, and each transformation is different. It is because of this great freedom of choice that we can transform a single problem into a multitude of others.

To illustrate how the transformation is carried out, we will choose a very simple function, z^2, and show what the corresponding transformation does to a simple diagram.

Consider the four squares (Figure 78), formed by parts of the lines $x = 1$, $x = 2$, $x = 3$, $y = 1$, $y = 2$, $y = 3$ on ordinary graph paper.

The transformation is carried out in three stages. (i) Corresponding to each point of the diagram we write down a complex number. (ii) We square that number. (This is because we chose z^2 as our function. If we had chosen some other function, $f(z)$, we should have calculated that function of our complex number.) (iii) We turn the number given by process (ii) back into a point on the diagram.

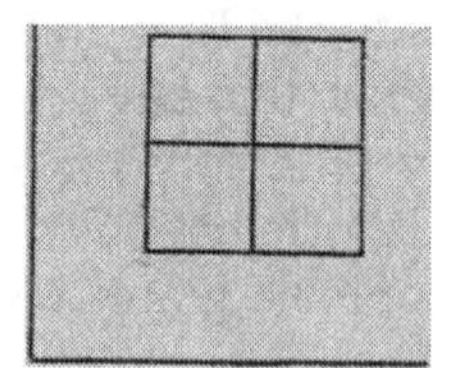

Figure 78

All of these processes are easy to carry out. For example, if we want to find where the transformation sends the point (2, 3), the calculations would run as follows.

Stage (i). Corresponding to the point (2, 3) we take the complex number $2 + 3i$. The general rule is that corresponding to any point (x, y) we take the complex number $x + iy$.

Stage (ii). We square the number found in Stage (i). The square of $2 + 3i$ is $-5 + 12i$.

Stage (iii). We find the point corresponding to $-5 + 12i$. It is $(-5, 12)$. The rule is the same as that used in the first stage, but working in the opposite direction.

The result of the whole calculation is that the point (2, 3) is transformed to the position $(-5, 12)$.

The same calculation is made for each of the nine points where the lines of the original diagram cross each other. In this way it is found that (1, 1) goes to (0, 2), (1, 2) goes to $(-3, 4)$, (1, 3) goes to $(-8, 6)$, and so on for the rest of the points.

We plot the nine new points on graph paper, and join them up by curves, corresponding to the lines of the original figure.

The result is shown in Figure 79 where we have the diagram obtained from the original diagram by the transformation z^2.

You will see that a transformation acts something like a distorting mirror. When you look at yourself in the bowl of a spoon or a polished jar, you see a creature with a face and hands that are reminiscent in a vague way of your own face and hands, yet without being faithful copies of the originals. Something in the way of proportions has been destroyed, yet something also has been preserved that enables you to recognize the reflection as the reflection of a human being.

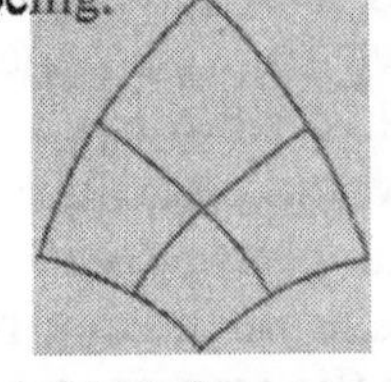

Figure 79

Conformal transformations distort straight lines into curves and alter distances. They preserve angles – in the last diagram above the curves still meet at right angles, as did the lines in the original diagram. And – the important property for practical applications – they preserve the quality of being a solution of a physical problem. If the original diagram represented lines of force or lines of flow (in a certain class of problems), so will the new diagram represent lines of force or lines of flow.

A very striking example of this principle is the Joukowski aerofoil.

A problem with a simple and elementary mathematical solution is to determine the way in which a stream passes round a circular obstacle. The lines of flow are something like the lines shown in Figure 80. This is our starting point, a problem the solution of which is known. Since a circle is such a simple shape, it is not surprising that this problem should have an exact, simple solution.

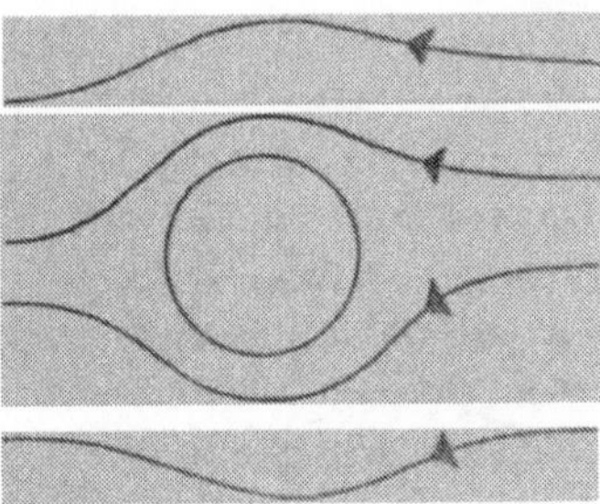

Figure 80

The practical problem is to find the lines of flow for a stream of air as it passes the wing of an aeroplane, the section of which is nothing so simple geometrically as a circle.

Joukowski discovered that, by choosing quite a simple function, namely $z + 1/z$, and applying the corresponding conformal transformation, the shape of the circle could be transformed into something which at any rate resembled the section of an aeroplane's wing. The curve to which the circle is transformed is shown in Figure 81. This curve is not used in practice for the section of any wing; it would be inefficient if it were. But the curve is useful theoretically as giving some indication of how an air-stream is likely to behave in passing round an obstacle of this general shape. Without the transformation, to cope with such a problem at all would be a very difficult matter.

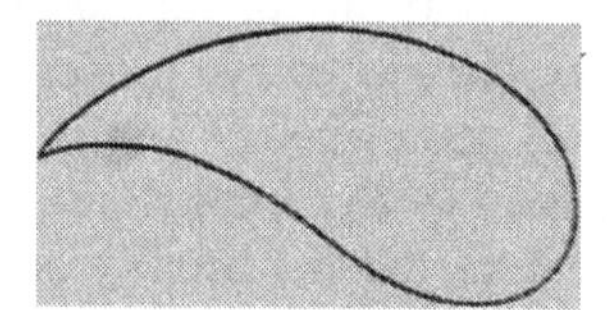

Figure 81

To solve the problem by means of the transformation, one simply transforms the whole diagram for the flow round a circle. The circle goes into the aerofoil. The lines of flow past the circle are transformed into lines of flow past the aerofoil.

The diagram above for the flow past a circle is simply a rough sketch. Anyone who is interested in the mathematical solution and an accurate diagram of the flow will find it in Lamb's *Hydrodynamics* (pages 77, 78). Joukowski's aerofoil, and other conformal transformations, are described in Phillips' *Functions of a Complex Variable.* To read these books a mathematical training is necessary.

CHAPTER THIRTEEN

Finite Arithmetics and Geometries

Curr, whom we have already quoted in connection with the Australian tribes, claims that most of these count by pairs. So strong indeed is this habit of the aborigine that he will rarely notice that two pins have been removed from a row of seven; he will, however, immediately become aware if one pin is removed.

Tobias Dantzig, *Number, the Language of Science*

The topics presented in this chapter are something in the nature of novelties, and hence interesting to the human mind. This is the aspect of them that first strikes one. They also have a certain value in various branches of higher mathematics. This is their second aspect. Their third aspect – a most unexpected one – is that of practical utility. Bizarre as they may seem, they can be useful. This chapter concludes with a brief account of one application of finite arithmetics – to agricultural research.

AN ALTERNATIVE ARITHMETIC

Few of us, I suppose, have ever verified directly by counting that $12 \times 12 \times 12$ is really 1,728. For a great many people arithmetic is taken on trust. Certain rules are taught in schools and these are not questioned. Even in this relatively enlightened century that probably applies to the majority of school children and school teachers. Others, of a more critical cast of mind, are reassured by the general consistency of arithmetic. Often there are several different ways of performing a calculation, but all lead to the same answer. For example, if we are asked to find $5 \times (7 + 3)$, we can either add the 7 and 3 to get 10, and then multiply by 5 to get 50, or we can write the expression as $(5 \times 7) + (5 \times 3)$, which equals $35 + 15$, again leading to the answer 50. The rules are at least consistent, and this suggests that they may be true.

The arithmetic about to be described, while different from the arithmetic in common use, would pass the tests applied above. It can be taught by rule, the rules resemble very closely those of ordinary arithmetic, and they are consistent, they always lead to the same answer.

Let us imagine a young child on entering school required to memorize the two tables given below, an addition table and a multiplication table. If it gives an answer other than that shown in these tables, it is beaten. Apart from these tables, the ordinary rules of arithmetic apply.

In this new arithmetic, there are only five numbers, 0, 1, 2, 3 and 4. No question ever leads to any answer other than one of these five. The child does not have to bother with tens and units columns, or anything of that sort.

Here are the two tables it has to memorize.

ADDITION TABLE

	0	1	2	3	4
0	0	1	2	3	4
1	1	2	3	4	0
2	2	3	4	0	1
3	3	4	0	1	2
4	4	0	1	2	3

MULTIPLICATION TABLE

	0	1	2	3	4
0	0	0	0	0	0
1	0	1	2	3	4
2	0	2	4	1	3
3	0	3	1	4	2
4	0	4	3	2	1

To find, for example $1 + 3$ the child looks at the addition table. It looks along the row opposite 1 and travels along this row until it is in the column underneath 3. There it finds the answer, 4 – quite in agreement with orthodox ideas on arithmetic. Less orthodox is the result $3 + 4 = 2$, and, in the multiplication table, $2 \times 3 = 1$. However, the child knows nothing of orthodox arithmetic. It learns these tables as it is told.

But suppose it wishes to check a calculation of the same type as we considered earlier, for example $3 \times (2 + 4)$. It may say, this is $(3 \times 2) + (3 \times 4)$. From the multiplication table, $3 \times 2 = 1$ and $3 \times 4 = 2$. So the answer is $1 + 2 = 3$. Or it may say, $2 + 4 = 1$, $3 \times 1 = 3$. Reaching the same answer by either route, it is satisfied with the correctness of what it has learnt.

You may like to check other calculations in this arithmetic. You will find that, if you use these tables correctly, you always come to the same answer, whatever route you follow.

Indeed, this arithmetic has many advantages over the ordinary one. For instance, there are no fractions or negative numbers in it. We have results like $2 \div 3 = 4$, $2 - 3 = 4$. Any number can be divided exactly by any number (except of course 0) without remainder. Every number has a reciprocal. The reciprocal of 1 is 1; of 2 is 3; of 3 is 2; of 4 is 4.

A question like

$$\text{Simplify } \frac{1\frac{1}{3}}{2\frac{1}{4}}$$

is worked as follows. $\frac{1}{3}$ is 2 (since $3 \times 2 = 1$). So $1\frac{1}{3} = 1 + 2 = 3$, the value of the numerator. $2\frac{1}{4}$ is found similarly. $\frac{1}{4} = 4$ (since $4 \times 4 = 1$). So $2\frac{1}{4} = 2 + 4 = 1$. The fraction is accordingly $\frac{3}{1}$, that is, 3.

An algebra can be built on this arithmetic and works very much like ordinary algebra. A quadratic equation can be solved by completing the square, and never has more than two solutions. For example, if we have to solve $x^2 - 2x + 2 = 0$ we add 3 to both sides, giving $x^2 - 2x = 3$. (Remember $2 + 3 = 0$.) To complete the square we add 1. $x^2 - 2x + 1 = 4$, that is, $(x-1)^2 = 4$. 4 is the square of 2, but it is also the square of 3 (see multiplication table). So $x - 1$ is either 2 or 3, that is x must be 3 or 4. You can check the correctness of these answers by substituting the answers in the original equation; of course you must be careful not to lapse into the addition and multiplication tables you learnt at school.

In such an algebra there would be a limit to the number of quadratic equations. The coefficient of x^2 could only be 1, 2, 3, 4; if it were 0 the equation would not be quadratic. The coefficient of x, and the constant term could each be chosen from the numbers 0, 1, 2, 3, 4. Altogether there would be a hundred quadratics, so it would be possible for a pupil to say that it had solved *all* possible quadratic equations.

CONSTRUCTION OF FINITE ARITHMETICS

Where do the tables given above come from? How are they made up?

I was once told of a bank where the clerks were alarmed to find that they were exactly a million pounds out in their accounts. For the truth of the story I cannot vouch. It was said that they were using a calculating machine, which only carried six figures for the pounds, so that the largest amount it could show was £999,999/19/11¾. It was round about this amount, and someone had added £5, with the result that the machine showed £000,004/19/11¾, exactly a million pounds less than it should have.

The arithmetic we have just considered would arise from a calculating machine which only carried the numbers 0, 1, 2, 3, 4 arranged in a circle, so that going a step beyond 4, 0 would again be reached. The effect of such a machine is that adding 5 makes no difference to it. No record is kept of how many times the wheel has turned. Every number is replaced by its remainder on division by 5. For example, $3 \times 4 = 12$. The remainder on dividing 12 by 5 is 2, so in the multiplication table 3×4 appears as 2.

The simplest arithmetic of this kind is that with only two numbers, 0 and 1, and the tables

ADDITION

	0	1
0	0	1
1	1	0

MULTIPLICATION

	0	1
0	0	0
1	0	1

This arithmetic arises if we replace every number by its remainder on being divided by 2. If you like, you can interpret 0 above as meaning 'Even', 1 as meaning 'Odd'. The addition table above then takes the perfectly intelligible form

Even + Even = Even
Even + Odd = Odd
Odd + Odd = Even,

and similarly for the multiplication table.

This seems to be the arithmetic of the tribe mentioned in the quotation at the head of this chapter. Apparently they count with their hands, rather than their fingers, 'Left, Right, Left, Right, Left, Right ...' In effect, 'Left' means 'Odd' and 'Right' means 'Even', with the curious consequence that they fail to detect the theft of an even number of articles.

In Chapter 7, under the heading 'The Algebra of Classes', we had occasion to use this arithmetic, with $1 + 1 = 0$. This is an example of an application of this arithmetic to another branch of mathematics.

FINITE GEOMETRIES

We are now ready to create yet another universe. In Chapter 11 the first clause stated, 'The universe is to contain points specified by numbers (X, Y, Z)'. These numbers were, of course, the numbers of ordinary arithmetic. The universe would therefore contain an infinite number of points. But we are now in a position to amend this clause so as to get a universe containing only a finite number of points. All we have to do is to say that the numbers X, Y, Z shall be numbers belonging to one of the finite arithmetics discussed in this chapter, and operated with according to the laws of that arithmetic.

The arithmetic containing only the numbers 0 and 1 is called 'the arithmetic modulo 2'. Let us see what sort of a universe we would get if we said that each point was to have a label (X, Y, Z), the numbers being drawn from the arithmetic modulo 2. As usual, we exclude (0, 0, 0). If (0, 0, 0) were allowed in, we should have 8 points altogether. Since it is excluded we have only 7. They are listed below.

$A\,(1, 0, 0)$	$C\,(0, 0, 1)$	$E\,(1, 0, 1)$	$G\,(1, 1, 1)$
$B\,(0, 1, 0)$	$D\,(0, 1, 1)$	$F\,(1, 1, 0)$	

These seven points constitute our universe.

In Chapter 11 we saw that all the points on a line could be obtained by 'mixing' two points. What can we get by mixing, say, A and B? We are very limited in our mixtures, because only the numbers 0 and 1 are at our disposal. The only conceivable mixtures are the following.

(i) 1 of A and 0 of B. This is simply A.

(ii) 0 of A and 1 of B. This is simply B.

(iii) 1 of A and 1 of B, that is, $A + B$. Adding the co-ordinates of A and B we have

$$\begin{array}{r} 1,\ 0,\ 0 \\ \underline{0,\ 1,\ 0} \\ \text{Total}\ \ 1,\ 1,\ 0, \end{array}$$ that is to say, F.

(iv) 0 of A and 0 of B. But this gives (0, 0, 0) which is not allowed.

Accordingly, there are only three points on the line AB, namely, A, B, and F.

Exactly the same argument shows that there are only three points on the line joining any two points; the points on AB were A, B, and $A + B$ which is F. In the same way, the points on AC are A, C and $A + C$, which is E.

Altogether we can find the lines listed below:

1. *BDC*	3. *AFB*	5. *BGE*	7. *DEF*
2. *AEC*	4. *AGD*	6. *CGF*	

Unfortunately, it is not possible to draw this universe on paper in such a way as to show the straight lines as actually being straight. Figure 82 makes the first six lines appear straight, but the seventh line, *DEF*, looks quite wrong. The curved dotted line indicates that D, E and F ought to be in line.

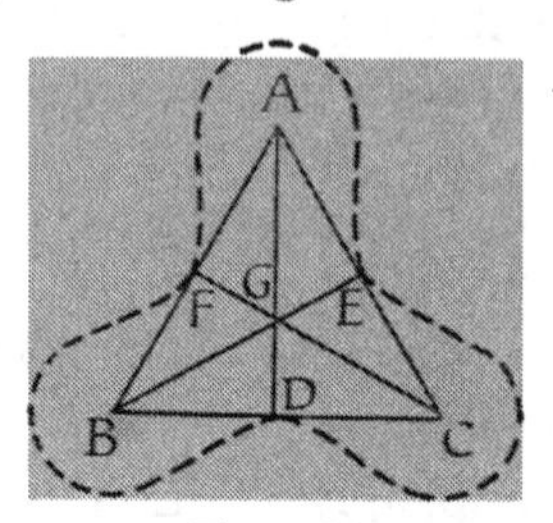

Figure 82

The equations of the seven lines listed above are

1. $X = 0$	3. $Z = 0$	5. $X + Z = 0$	7. $X + Y + Z = 0$
2. $Y = 0$	4. $Y + Z = 0$	6. $X + Y = 0$	

In checking these equations, you must not forget that $1 + 1 = 0$. Indeed, this must be borne in mind throughout the work.

The diagram above is very unsatisfactory, because it suggests that there are differences between the points. G, for instance, is in the centre of the diagram. This is true of no other point. The line *DEF* appears curved, while the other lines appear straight.

Actually this is an extremely democratic universe. Each point is just as good as every other point; each line as every other line.

For example, on the line AB you may think, from the diagram, that F is between A and B, as suggested by the equation $F = A + B$. But the relation between A, B and F is perfectly symmetrical. It is equally true that $A = F + B$ and that $B = F + A$. Each point is a mixture of the other two, as you can verify by adding the co-ordinates.

In Chapter 10, in connexion with Desargues' Theorem, we discussed diagrams in which it was possible to re-letter the figure in many ways without destroying the truth of the theorem. The diagram above has the same property. There are altogether 168 ways of putting the letters on it. As it stands, the letters *A*, *B*, *C* are at the corners of the triangle. You can re-letter the diagram in such a way that any three letters you like come at the corners, provided only that these letters do not belong to three points in line with each other. You could, for example, put *E*, *F*, *G* at the corners; but you could not put *D*, *E*, *F* at the corners, because *DEF* is line 7.

The universe just described is of interest in connexion with the theory of groups. From it we can also obtain the idea of having matrices built up, not from ordinary numbers, but from numbers in a finite arithmetic.

AGRICULTURAL RESEARCH

Suppose it is desired to test a number of varieties of wheat. At first sight nothing seems easier. Plant some of each kind in a field, and see which does best. Perhaps the wheat sown in the North-East corner does best. But then the objection may be raised – perhaps the North-East corner was a more favourable position for wheat than any other part of the field, perhaps the soil there is richer, perhaps the good crop has nothing at all to do with the variety of wheat, but only with where it was planted. In short, what one crop does is not evidence.

The way to meet this objection is obvious. The same variety of wheat must be planted in all kinds of different situations, so as to eliminate, as far as possible, the effect of differences in soil fertility. Scientific experiments of this sort are usually subjected to statistical analysis, to show how much weight can be attached to the results of the experiment, and what chance there is that the apparent result is purely due to outside causes. Accordingly, it is necessary to have some systematic way of mixing up the various types of wheat, so as to make the statistical analysis simple to carry out.

Such a system can be worked out with the help of finite arithmetic. We will use the arithmetic explained at the beginning of this chapter, that is, the arithmetic modulo 5, with the numbers

0, 1, 2, 3 and 4. With the help of this arithmetic we will obtain a scheme for the planting of four square plots with five varieties of wheat, that are to be compared with each other.

We begin by setting out the numbers in the following way.

0 1 2 3 4	0 1 2 3 4	0 1 2 3 4	0 1 2 3 4
1	2	3	4
2	4	1	3
3	1	4	2
4	3	2	1

The row at the top is the same in each case, 0, 1, 2, 3, 4. The columns are taken from the multiplication table. When the numbers 0, 1, 2, 3, 4 are multiplied by 1 we get 0, 1, 2, 3, 4; these numbers form the first column. When 0, 1, 2, 3, 4 are multiplied by 2 we get 0, 2, 4, 1, 3; these numbers form the second column above. The third column, in the same way, is found from multiplication by 3; the last column from multiplication by 4.

The squares are then completed by means of the addition table. For instance, in the second square

0 1 2 3 4
2   :
4   :
1 4
3

4 is written in the space shown because it is the sum of 1 and 3. In order to fill in any space, we look at the number at the beginning of the row containing that space, and the number at the head of the column containing that space; the sum of these two numbers is then written in. By the 'sum' of course we understand the sum in the arithmetic modulo 5, so that $4 + 3$ means 2.

In this way we arrive at the four squares shown below

0 1 2 3 4	0 1 2 3 4	0 1 2 3 4	0 1 2 3 4
1 2 3 4 0	2 3 4 0 1	3 4 0 1 2	4 0 1 2 3
2 3 4 0 1	4 0 1 2 3	1 2 3 4 0	3 4 0 1 2
3 4 0 1 2	1 2 3 4 0	4 0 1 2 3	2 3 4 0 1
4 0 1 2 3	3 4 0 1 2	2 3 4 0 1	1 2 3 4 0

This gives us our pattern for planting the wheat; wherever 1 appears, we plant the first variety; wherever 2 appears, the second, and so on; where 0 appears, we plant the fifth variety.

If you examine any one square – the first for example – you will see that there is a 0 in each row, so that, if the field happened to get gradually more fertile as one went from North to South, it would not affect the average scored by variety 0. There is one 0 in each column; so that a gradual change of fertility in the East-West direction would not affect the result. What is true of 0 is true of the other four numbers.

Nor is there any harmful connexion between the layout in one square and in the next. In the first square 0 occurs in the positions

```
* . . . .
. . . . *
. . . * .
. . * . .
. * . . .
```

Look at the second square, and see what numbers occur in these positions. They are 0, 1, 2, 3, 4 – each number once. You will find a similar result for any square and any number. Pick any square you like, and note the positions in which any particular number occurs. Go to any other square, and note the numbers that occupy these positions. You will find the positions are shared out equally between all the five varieties.

The arrangements above are known as a set of 'orthogonal Latin squares'.[1] The finding of such sets looks like a purely trivial puzzle; in fact it is an important practical task.

In a problem of applied mathematics, it is often helpful to begin by searching through pure mathematics, and seeing if there may not be somewhere in the literature a procedure which has already provided the pattern required for the solution of the practical problem.

EXTENSIONS OF FINITE ARITHMETICS

In the arithmetic we have just made use of, there were exactly five quantities 0, 1, 2, 3, 4. We can enlarge this arithmetic in the following way.

In that arithmetic, the equation $x^2 = 1$ has the solutions 1 and 4, the equation $x^2 = 4$ has the solutions 2 and 3, and the equation $x^2 = 0$ has the solution 0. The equations $x^2 = 2$ and $x^2 = 3$ have no solutions.

1. Mann, *Analysis and Design of Experiments.*

We can bring in a new symbol, $\sqrt{2}$, to provide a solution for the equation $x^2 = 2$.

We now have at our disposal the following 25 quantities

0	1	2	3	4
$\sqrt{2}$,	$\sqrt{2}+1$,	$\sqrt{2}+2$,	$\sqrt{2}+3$,	$\sqrt{2}+4$
$2\sqrt{2}$,	$2\sqrt{2}+1$,	$2\sqrt{2}+2$,	$2\sqrt{2}+3$,	$2\sqrt{2}+4$
$3\sqrt{2}$,	$3\sqrt{2}+1$,	$3\sqrt{2}+2$,	$3\sqrt{2}+3$,	$3\sqrt{2}+4$
$4\sqrt{2}$,	$4\sqrt{2}+1$,	$4\sqrt{2}+2$,	$4\sqrt{2}+3$,	$4\sqrt{2}+4$

These again form a finite arithmetic; you can add, subtract, multiply and divide (except by 0) without ever going outside the 25 symbols above.

The theory of such extensions is known as the theory of Galois Fields. This theory belongs to what is called Modern Higher Algebra, which sounds imposing. Galois Fields however, since they consist of only a finite number of elements, are a particularly simple and easy branch of modern algebra.

An interesting property of the set of numbers above is that if one takes $\sqrt{2}+2$ and keeps multiplying it by itself one obtains in turn all the numbers of the set, except 0. That is to say, every number but zero is a power of $(\sqrt{2}+2)$ and can be written as $(\sqrt{2}+2)^n$. for some whole number n. A property of this kind holds for every Galois Field.[1]

1. The word 'Field' has a special meaning in algebra. It has nothing whatever to do with the fields of wheat mentioned earlier, but indicates a set of symbols within which one can add, subtract, multiply and divide – what I have called 'an arithmetic'.

CHAPTER FOURTEEN

On Groups

> The mathematics of the twenty-first century may be very different from our own; perhaps the schoolboy will begin algebra with the theory of substitution groups, as he might now but for inherited habits.
>
> *Simon Newcomb*, 1893

Towards the end of Chapter 7 various groups of movements were discussed. The arguments of that chapter were, I think, simple ones. The main difficulty the student of groups meets is not that of following the argument, which is nearly always straightforward, but of grasping the purpose of the investigation. This chapter tries to deal, to a limited extent, with the question, 'How did the theory of groups arise, and what is it for?'

THE AXIOMS OF A GROUP

In Chapter 7, although we used the word 'group', we did not say exactly what it implied.

The first thing we understand by the use of the word group is that we are dealing with a collection of symbols, operations or things that can be combined in some way, and that when combined still give something belonging to the collection. For example, when we multiply two numbers together (which is a way of combining the two numbers) we expect the result to be a number. However little we know about arithmetic, we should be surprised if the answer was a bunch of parsley. In the same way, movements can be combined; a rotation through 90° followed by a rotation through 45° gives a rotation through 135°; the combined effect of two movements is again a movement.

Group theory only gradually emerged as a precise theory; in the early days the above property was the only one emphasized. It was often referred to as 'the group property'.

To-day, there are three other requirements before a collection of symbols can be called a group.

(i) The collection must contain a symbol I which has no effect on any of the other symbols when it is combined with them. That is to say, if X is any other symbol, we must have $I.X = X$, and $X.I = X$.

(ii) Every symbol must have an inverse, that is to say, whatever a symbol does, another symbol must exist that undoes it. For example, if a group contains the operation 'multiply by 2' it must also contain the operation 'multiply by $\frac{1}{2}$'. So for example, the collection of operations consisting of 'multiplication by whole numbers' do not form a group.

(iii) The symbols must obey the Associative Law. That is, if X, Y, Z are any symbols in the collection, $(XY)Z$ and $X(YZ)$ must mean the same thing. In any group consisting of operations this law is automatically satisfied. Suppose, for example, the operations are movements. Suppose Z shifts an object from position (1) to (2), while Y shifts it from (2) to (3) and X shifts it from (3) to (4). XY means the combined effect of movement Y followed by movement X (remember the reversal of order in writing), i.e. XY shifts from (2) to (4). $(XY)Z$ means that we do operation Z, and then operation XY as just specified. Z shifts from (1) to (2) and XY from (2) to (4); the combined effect is to take the object from (1) to (4). In the same way, you can see that $X(YZ)$ also means (1) to (4).

All this amounts to, in fact, is that saying, '(1) to (2)' and then *in one breath* 'and–(2)–to–(3)–to–(4)' is no different in its meaning from saying breathlessly '(1)–to–(2)–to–(3)' and then calmly 'and (3) to (4)'.

The first statement corresponds to $(XY)Z$; the second to $X(YZ)$, as you can read off from Figure 83.

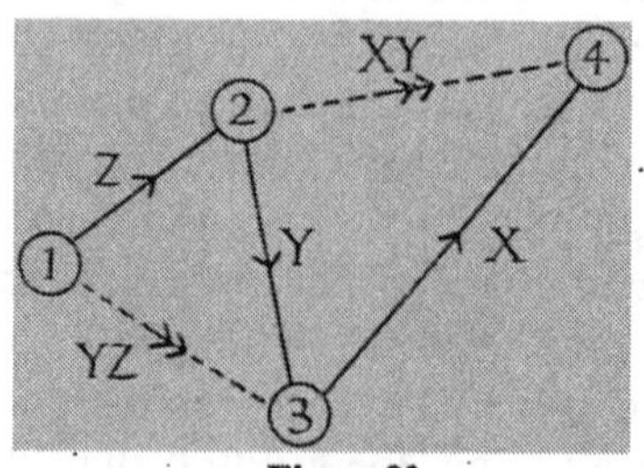

Figure 83

As we shall be concerned mainly with groups of operations, you need not pay much attention to requirement (iii).

A group of operations then has the properties that any two operations combined are equivalent to an operation of the set; there is an operation, I, that consists of leaving things just as they are; and whatever you do, you can also undo.

What is very remarkable is the immense theory that can be built on these few assumptions.

You may like to turn back to Chapter 7, and check that these assumptions do hold for the group of the rectangle and the group of the equilateral triangle. In the group of the rectangle what is the operation that undoes p? p itself.

THE ORIGIN OF GROUP THEORY

The need for group theory arose in connexion with a question which, in itself, is of no practical importance, but as a model for a method of investigation is of the utmost importance in all branches of mathematics.

The problem was that of solving algebraic equations. Linear and quadratic equations were solved in ancient times. Equations of the third and fourth degree were solved shortly before 1550. And there things stuck. Many mathematicians tried to solve the equation of the fifth degree, but none succeeded. Some unification was achieved; it was shown that all the methods that had worked with equations of the 1st, 2nd, 3rd and 4th degree could be regarded as particular cases of one single method, and *this method failed when applied to the equation of the fifth degree.* To solve a problem means to reduce it to something simpler than itself. It gradually began to occur to people that perhaps the equation of the fifth degree could not be reduced to anything simpler than itself. Eventually this was proved to be the case.

ATOMIC PROBLEMS

The equation of the fifth degree therefore appeared as a kind of mathematical atom – something that could not be split into anything simpler by algebraic means.

A very simple example will serve to illustrate this idea of an atomic problem. If we want to solve the equation $x^6 = 2$, we can, if we like, do it in two steps. With a table of square roots we can find $\sqrt{2} = 1{\cdot}4142$, and with a table of cube roots we find that the cube root of 1·4142 is 1·1223. A table of sixth roots is therefore not essential; a sixth root can be found with the help of tables of square and cube roots. Algebraically this means that solving the equation $x^6 = 2$ may be replaced by solving the equations $y^2 = 2$, $x^3 = y$.

Accordingly, solving $x^6 = 2$ is not an atomic problem; it can be split up into the solving of two simpler problems.

On the other hand $x^2 = 2$ cannot be dealt with in this way. This problem is atomic. It cannot be broken up into two simpler problems.

With any algebraic equation one may therefore ask whether it can be split up into two simpler equations. If it cannot, it is (so far as algebra is concerned) atomic, and it is a waste of time looking for tricks to solve it. It is noteworthy that mathematicians passed more than 250 years looking for tricks to solve the fifth degree equation, before they realized they were attempting the impossible.

An atomic problem is not absolutely insoluble. If one brings up heavier artillery, it may cease to be so. For instance, square roots can be found with the aid of logarithms. A problem is only relatively insoluble; insoluble with given tools – like the wooden ploughs of Chapter 5. A great saving of time is effected by a theory which shows the impossibility of solving a problem by a given type of method. One then sets out straight away to find a new kind of method, instead of passing a century or so looking for tricks within the old methods.

Galois, a brilliant mathematician who was killed at the age of twenty-one (in 1832), showed that every algebraic equation was connected with a group, and by examining this group one could say whether the equation was atomic or not. If the group was what is called *compound*, the equation could be broken up into two simpler equations; if the group was *simple*, the equation was atomic.

The Galois Theory takes a certain amount of mastering. In works on modern higher algebra it is presented in a form rather different from that in which Galois left it.

Here I cannot do more than try to give a very faint idea, by one or two illustrations, of how an equation has a group associated with it, and of what is meant by the group being compound or simple. These illustrations may have something of the appearance of a conjuring trick; I have not space to show the theory which leads me to these particular examples.

A CYCLIC EQUATION

First of all we consider the equation $x^3 - 3x + 1 = 0$, the roots of which we will call a, b and c. As a matter of fact, these roots can be given by means of trigonometry; a is 2 cos 40°, b is 2 cos 80°, and c is 2 cos 160°. If you look at the three angles here, you will see that 80° is twice 40°, and 160° is twice 80°. This suggests that we ought to see what we get if we double the angle again. Twice 160° is 320°, and the cosine of 320° is the same as the cosine of 40°, so we have here the pattern of the snake biting its own tail. The operation (doubling the angle) that leads us from a to b, also leads us from b to c and from c to a. The letters a, b, c exhibit the same kind of symmetry as the toothed wheel shown here.

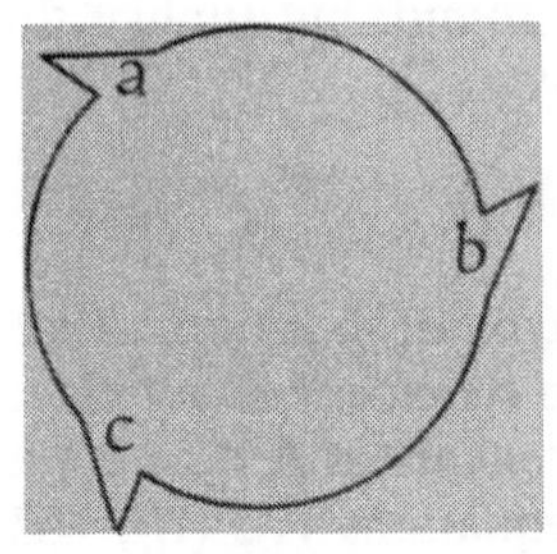

Figure 84

We could turn this wheel in such a way that where a previously stood, we saw b; where b was, we saw c; where c was, we saw a.

The relationship of a, b, and c can also be shown by means of the equations

$$\text{(I)}\quad b = a^2 - 2$$
$$\text{(II)}\quad c = b^2 - 2$$
$$\text{(III)}\quad a = c^2 - 2.$$

The symmetry of these relations is worth considering. Certainly these equations are fair to a, b and c; no letter can complain of being unfairly treated. And yet there is not total symmetry. For instance, in equation (I) we cannot exchange a and b without producing an untrue statement. The actual relation of b to a is given by the equation

$$\text{(IV)}\quad a = -b^2 - b + 2$$

which of course has its companion equations

$$\text{(V)}\quad b = -c^2 - c + 2$$
$$\text{(VI)}\quad c = -a^2 - a + 2$$

One might compare it with the situation that sometimes arises in games, B can beat A, C can beat B, A can beat C.[1] This situation is symmetrical, but the symmetry has, so to speak, a twist in it. The companion statements 'A loses to B', 'B loses to C', 'C loses to A' have the same symmetry.

The situation with complete symmetry is when B draws with A, C with B, and A with C. One can then interchange A and B in the statement 'B draws with A' to obtain the equally true statement 'A draws with B'.

Complete symmetry corresponds to the group of the equilateral triangle, discussed at the end of Chapter 7.

The type of symmetry that is possessed by the toothed wheel, by the equations (I), (II), (III), and by the players when A loses to B, B to C, and C to A, is called *cyclic symmetry*. The toothed wheel can be turned round, but (unlike the equilateral triangle) it cannot be turned over.

THE CYCLIC GROUP

When we have this type of symmetry, that of a dog chasing its tail, it is clear that if a statement about players A, B, C is true, the corresponding statement about B, C, A will be equally true. In the same way, if from equations (I), (II), (III) we can deduce an equation $f(a, b, c) = 0$, we shall also be able to deduce an equation $f(b, c, a) = 0$.

By what procedure is the second equation obtained from the first? Wherever a occurs in the first equation, we rub it out and write b; where b occurs, we rub it out and write c; where c occurs, we rub it out and write a.

The operation just explained is often denoted by the symbol (abc). Each letter becomes the letter that follows it inside the bracket; a becomes b, b becomes c. Now c is the last letter in the bracket and strictly speaking, nothing follows it; we, however, agree to regard the first letter in the bracket as coming after the last; thus c becomes a.

For short, I shall use the letter w to denote (abc). Thus, if we

1. An African schoolboy once asked me how, in view of the possibility of this situation, one could maintain that the statement 'If x is greater than y, and if y is greater than z, then x is greater than z' was correct. As with most questions asked me by children, I found I had to think very hard before I could answer it.

have any function $f(a, b, c)$, then $w.f(a, b, c) = f(b, c, a)$. Read this as, 'The operation w acting on $f(a, b, c)$ gives $f(b, c, a)$'.

This operation could of course be repeated. Applying it again we find $w^2.f(a, b, c) = f(c, a, b)$. And it could be applied yet again, to give $w^3f(a, b, c) = f(a, b, c)$. Thus the operation w applied three times brings us back to where we started, and we may write $w^3 = I$.

The operations I, w, w^2 form a group, technically known as 'the cyclic group of order 3', or C_3 for short. This group is the group associated with the equation $x^3 - 3x + 1 = 0$, and is referred to as the Galois Group of the equation.

If we have any true relationship involving the roots a, b, c of this equation, and we apply to this relationship any operation of the Galois Group, we shall obtain a true relationship as the result. For example, equation (IV) given earlier in this section is a true relationship between a, b and c. If we apply the operation I, that is to say, if we leave it alone, we simply have equation (IV). If we apply the operation w to equation (IV), we obtain equation (V), which is of course a true relationship between the roots. If we apply operation w^2, we obtain equation (VI).

I expect that, when you were reading the earlier part of this section, and you came to equation (IV), you felt that it was reasonable that it should have the companion equations (V) and (VI), and I expect you could have written these equations down yourself if I had not given them. So do not regard my explanation of the operations w and w^2 as something new; *the operations* w *and* w^2 *are simply what you were doing when you saw that equations* (V) *and* (VI) *were the natural companions of equation* (IV). The object of bringing in these operations w and w^2 is to replace the vague sense of pattern with which we started by a precise and conscious idea, which can be expressed by means of symbols and form the foundation for a mathematical theory.

The Galois Group consists of those interchanges which can be made between the roots of an equation, such that any statement, which is true before the interchanges are made, will give a true statement after the interchange.

A RESTRICTION ON STATEMENTS

We have to be a little careful here about what we mean by 'statement'. We said earlier that, for the three players A, B, C

in the circumstances then considered, any statement that held about A, B, C would hold equally well for B, C, A. But this clearly does not apply to such a statement as 'A has a red nose'; we cannot deduce from it that B and C also have red noses. The statement must be about the game. Nor will any statement about the game do. For example, from the statement 'A has a strong forehand drive' it does not follow that B, who beats A, also has a strong forehand drive. B's victory may be due to craftiness, which gives A no chance to use his forehand drive. In fact, the statements must be confined to statements about the results of the game; any statement probing too deeply into how that result is achieved must be disqualified for our present purposes.

A similar limitation is necessary in our algebraic problem. Consider, for example, the fact that a, which is $2 \cos 40°$, has the value 1·53208 ... If to the equation $a = 1{\cdot}53208$ ... we apply the operation w, then a is replaced by b and the equation becomes $b = 1{\cdot}53208$ ..., *which is not true.* We must restrict our functions $f(a, b, c)$ to polynomials in a, b, c with *whole numbers only* for coefficients. Thus an equation like $ac - abc + a^2 + a - 2 = 0$ qualifies all right as being 'a statement'; but we exclude expressions such as $a + b\sqrt{2}$ or $c - \pi$ in which irrational numbers, $\sqrt{2}$ and π, occur. Fractions, however, do no harm. For instance $\frac{1}{2}(a^2 + a + c) - 1 = 0$ means exactly the same as

$$a^2 + a + c - 2 = 0,$$

and is permitted to count as a statement. It is, in fact, equation (VI).

STATEMENTS ABOUT $\sqrt{2}$

Still using 'statements' in the same sense, let us consider what statements can be made about $\sqrt{2}$. Such a statement will take the form $f(\sqrt{2}) = 0$; or, if we prefer, $f(x) = 0$ for $x = \sqrt{2}$. $f(x)$ is to be a polynomial with whole number coefficients. Suppose we divide $f(x)$, whatever it is, by $x^2 - 2$. This will give us a quotient, say $\varphi(x)$, and a linear remainder, say $px + q$, where p and q will be whole numbers; you can see this is so, if you write down any polynomial with whole number coefficients, say

$$3x^4 - 7x^3 + 4x^2 + 9x + 5$$

and divide it by $x^2 - 2$. Nowhere in the long division is there any occasion for fractions to arise. When 23 is divided by 7 the quotient is 3 and the remainder 2; all of this is summed up in the single equation $23 = 7 \times 3 + 2$. In the same way, the division specified above can be summarized by the equation

$$f(x) = (x^2 - 2)\varphi(x) + px + q.$$

In this equation, we may put any value for x we like. Let us put $x = \sqrt{2}$. That will make $x^2 - 2$ become zero. Also, since $f(\sqrt{2}) = 0$, it will make $f(x)$ become zero. The equation above thus reduces to

$$0 = p\sqrt{2} + q$$

which seems to lead to the result $\sqrt{2} = -q/p$. Now it is well known that $\sqrt{2}$ cannot be expressed as a rational fraction, so we seem to have been led to a contradiction. The only escape is provided by the possibility that p and q are both zero. As this is the only way of avoiding an absurdity, it must be what happens. Accordingly $p = 0$, $q = 0$, that is to say, the remainder $px + q$ is zero. This means that the function $f(x)$ must divide exactly by $x^2 - 2$; we must have $f(x) = (x^2 - 2)\varphi(x)$.

What we have found is that we can make a statement, $f(\sqrt{2}) = 0$, only if $f(x)$ is of the form $(x^2 - 2)\varphi(x)$.

But now, if this is so, on putting $x = -\sqrt{2}$, we shall find $f(-\sqrt{2}) = 0$.

This means; *if any statement can be made about* $\sqrt{2}$, *the same statement can be made about* $-\sqrt{2}$. 'Statement' here must of course be understood in the special restricted sense we have been using; the above result would make nonsense if applied to a sentence like '$\sqrt{2}$ is a positive number'.

The proper use of the principle would be, for example, to deduce from the fact that $x = \sqrt{2}$ satisfies $x^3 - x^2 - 2x + 2 = 0$ that $x = -\sqrt{2}$ also satisfies it, since *an equation of this type* ('a statement') that holds for $\sqrt{2}$ must hold equally well for $-\sqrt{2}$.

THE EQUATION $x^4 - 10x^2 + 1 = 0$

The principle we have just had for $\sqrt{2}$ can be extended to give us the Galois Group of the equation $x^4 - 10x^2 + 1 = 0$, which has the four roots a, b, c, d where $a = \sqrt{2} + \sqrt{3}$, $b = \sqrt{2} - \sqrt{3}$, $c = -\sqrt{2} + \sqrt{3}$, $d = -\sqrt{2} - \sqrt{3}$. I will not go into the details of the proof, but the principle still holds that any statement about a, b, c, d will remain true if we everywhere replace $\sqrt{2}$ by

$-\sqrt{2}$. Now the effect of replacing $\sqrt{2}$ by $-\sqrt{2}$ is that a will change from $\sqrt{2}+\sqrt{3}$ to $-\sqrt{2}+\sqrt{3}$, i.e. it will become c. In the same way, c will become a. Further b will become d and d will become b.

The effect then of replacing $\sqrt{2}$ by $-\sqrt{2}$ is that a and c will change places, as will b and d. That is, $f(a, b, c, d)$ will become $f(c, d, a, b)$.

Now it is also true that no statement can be made about $\sqrt{3}$ that does not equally well apply to $-\sqrt{3}$. We can therefore change $\sqrt{3}$ into $-\sqrt{3}$. This has the effect of interchanging a and b, and also c and d. $f(a, b, c, d)$ becomes $f(b, a, d, c)$.

Finally, if we like, we may do both things simultaneously. We may replace $\sqrt{2}$ by $-\sqrt{2}$ and $\sqrt{3}$ by $-\sqrt{3}$. This will turn a into d, b into c, c into b and d into a. So $f(a, b, c, d)$ will become $f(d, c, b, a)$.

The operations listed above, together with I, the operation of leaving a, b, c, d exactly as they were, constitute the Galois Group of the equation. If $f(a, b, c, d) = 0$ is a true statement about the roots of this equation, so also will $f(c, d, a, b) = 0$, $f(b, a, d, c) = 0$, and $f(d, c, b, a) = 0$ be true statements.

The Group we have just found is one that we have met before. If the letters a, b, c, d are written on the corners of the rectangle in Figure 85, the interchanges of these letters found for the equation $x^4 - 10x^2 + 1 = 0$ are precisely the interchanges that arise if the rectangle is picked up and put back into its box in a different position. The group of the rectangle was the first group considered in Chapter 7.

Figure 85

Both with the equation $x^4 - 10x^2 + 1 = 0$, just considered, and with the equation $x^3 - 3x + 1 = 0$ considered earlier, we knew the roots of the equation before we knew the Galois Group. However, it is possible, by a procedure which is fairly complicated, to determine the Galois Group before the equation has been solved. If it were not so, the theory would be useless, for the whole point of the Galois Group is that it tells us *how hard it is going to be to solve the equation, and by what steps we should proceed.*

SIMPLE AND COMPOUND GROUPS

How does the group of an equation give us information about solving the equation? I will answer this question in one respect only, by showing how a group tells us whether a problem is atomic or not, that is to say, whether or not it can be broken up into simpler problems. I shall simply explain what one looks for in the group, without attempting to prove what I assert.

Let us consider an equation which has for its Galois Group a group mentioned in Chapter 7, namely, the group of the equilateral triangle.

An equation which has this group is the equation $x^3 = 2$. You might at first think this equation to be atomic. There does not seem to be any way of breaking up the extraction of a cube root. But you must remember that a cubic equation has three roots. The full solution of this equation is given by the roots a, b, c where

$$a = \sqrt[3]{2},$$

$$b = \sqrt[3]{2}\left\{\frac{-1+\sqrt{-3}}{2}\right\},$$

$$c = \sqrt[3]{2}\left\{\frac{-1-\sqrt{-3}}{2}\right\}.$$

Accordingly solving the equation fully involves finding two distinct things, namely $\sqrt[3]{2}$ and $\sqrt{-3}$. The problem therefore breaks up into two distinct problems, and is not atomic.

How does this capacity for being broken up show itself in the Galois Group of the equation? As was stated above, the Galois group is the group of the equilateral triangle, that is to say, the group with the table

	I	ω	ω^2	p	q	r
I	I	ω	ω^2	p	q	r
ω	ω	ω^2	I	r	p	q
ω^2	ω^2	I	ω	q	r	p
p	p	q	r	I	ω	ω^2
q	q	r	p	ω^2	I	ω
r	r	p	q	ω	ω^2	I

At a glance, this table shows a peculiarity. It breaks up into squares. The letters p, q, r occur in the North-East and South-West; the symbols I, ω, ω^2 in the North-West and South-East. If we were to write I, ω, ω^2 in red ink, and p, q, r in black ink, the table would show vividly how these letters occurred in blocks. If we were to look at the table from such a distance that we could distinguish only the *colours*, but not the actual letters, the table would appear to us as below

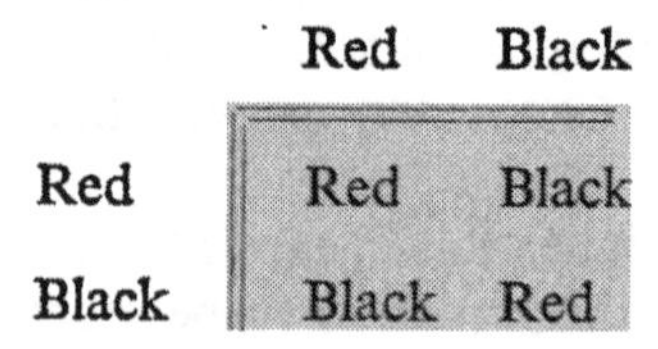

	Red	Black
Red	Red	Black
Black	Black	Red

Now this is itself the pattern of a group, a smaller group than the original one. We can read it as 'Red × Red = Red, Red × Black = Black', etc. This little group has the same pattern as the little group given in the last section of Chapter 7

	I	k
I	I	k
k	k	I

or as the multiplication table for the numbers 1 and –1

×	1	–1
1	1	–1
–1	–1	1

or as the addition table for Even and Odd

+	Even	Odd
Even	Even	Odd
Odd	Odd	Even

The bigger group, with the six elements, may thus be regarded as an elaboration of the pattern of the smaller group. We may say if we like that the big group is *mapped* on the small one by means of the correspondence

When an equation can be broken up into simpler equations, this fact always betrays itself in the pattern of the group table, which conceals within itself the pattern of a smaller group.

We had earlier in this chapter the equation $x^4 - 10x^2 + 1 = 0$, which, since it could be solved with the help of the two quantities $\sqrt{2}$ and $\sqrt{3}$, was not atomic. The Galois Group of this equation was the group of the rectangle, with the table

	I	p	q	r
I	I	p	q	r
p	p	I	r	q
q	q	r	I	p
r	r	q	p	I

The fact that the equation is not atomic is immediately seen if I and p are written in red, q and r in black.

When a group carries in this way within itself the pattern of a smaller group it is said to be *compound.* When it does not do so, it is called *simple.* Simple groups correspond to atomic problems.

The group I, w, w², that we met in connexion with the cyclic equation $x^3 - 3x + 1 = 0$, is simple. It contains 3 elements, and 3 is a prime number. Any group containing a prime number of elements is necessarily simple. Simple groups, which have a non-prime number of elements, are interesting and somewhat rare. A group known as the Group of the Icosahedron, with 60 elements, is the smallest such group. The insolubility of the equation of the fifth degree is connected with it. Next comes a group with 168 elements, which is connected with the finite geometry described in Chapter 13.

ANALOGIES OF THE GALOIS THEORY

The Galois Theory is interesting, but is it useful? The answer to this question provides a good illustration of the kind of usefulness a mathematical theory may have. For the Galois Theory has no direct application to any practical purpose. Nobody in practice wants to solve equations of the fifth degree. If an equation of the fifth degree arose in any technical activity, one would simply draw a graph, and see where the graph crossed the x-axis.

The real value of the Galois Theory is that it provides a model for almost any kind of investigation. The mathematician of older times asked, 'Can I find a trick to solve this problem?' If he could not find a trick today, he looked for one tomorrow. But the Galois Theory being known, we no longer assume that a trick need exist at all. We ask rather, 'Is there any reason to suppose that this problem can be solved with the means we have at hand? Can it be broken up into simpler problems? What is it that makes a problem soluble, and how can we test for solubility?' We no longer try to invent; we try to discover the nature of the problem we are dealing with.

Unlike the solution of algebraic equations, the solution of differential equations is a matter of great practical importance. Between 1883 and 1892 Picard and Vessiot successfully constructed a theory of differential equations, which very closely resembles, and was obviously inspired by, the Galois theory of algebraic equations.

An obvious task for mathematical research is to extend the ideas of Galois to cover all types of mathematical problem; to show which problems are atomic, and by what means compound problems may be recognized.

The usefulness of any such theory is, in the main, for the research mathematician. It saves him wasting his time trying to solve a problem by means which are inadequate to the task. It may suggest a systematic way of attacking new problems. The engineer rarely has time for fundamental mathematical research, and is generally content to use the methods developed by mathematicians, except in so far as he adapts them to the needs of practical calculation. The Galois theory has no direct application to technology.

Other Pelicans are described on the next few pages

Sir Charles Sherrington

MAN ON HIS NATURE

A322

This book, which was originally given as the Gifford Lectures, 1937–8, is a free, fearless, and invigorating expression of a biologist's philosophy. Sir Charles Sherrington contrasts a modern biologist's attitude towards the origins of life with that essentially held by a physician-philosopher of the sixteenth century. He shows that there is no longer any logical division of matter into living and non-living, and that 'growth' is likely to be expressible in terms identical with those used in atomic physics; chemical action, upon which growth depends, being but a name for a complex of inter-atomic electrical changes.

The principle of life, it may turn out, is no more than a useful human convention. But what of the Mind? Mind knows itself and knows the world. Chemistry and physics, explaining so much, cannot undertake to explain Mind itself. It can intensify knowledge of Nature, but it cannot be shown that Mind has hitherto directed the operations of Nature. In that sense Mind and Nature are different. Mind, by finding that 'blind' electrical-chemical forces are the instruments by which all matter exists, points to itself as something unique in Nature. And this the author accepts as a heartening challenge: if blind forces can do so much to wonder at, what cannot directed forces achieve?

2s 6d

Not for sale in the U.S.A. or Canada

SOME NEW PELICANS

A Short History of Confucian Philosophy – Wu-Chi Liu

A book for the general reader who wants to know at first hand about China's greatest philosophy, which has moulded the Chinese nation for almost twenty-five centuries. (A 333) *2s 6d*

How Money is Managed – Paul Einzig

The ends and means of monetary policy: the methods by which governments influence the movements of prices and other topics explained in terms which are readily understood. (A 312) *2s 6d*

Man, Morals and Society – J. C. Flugel

'Those who wish to know what psycho-analysis has to say on fundamental moral problems will here find an exposition written with great clarity and candour, based on a thorough grasp of all the relevant data and likely to stimulate further inquiry.' *The Spectator*. (A 324)* *3s 6d*

The Colour Problem – A. H. Richmond

A study of colour prejudice, racial discrimination, and social separation, with an account of racial relations and the 'colour-bar' in Britain and Commonwealth territories in Africa and the West Indies. (A 328) *3s 6d*

Sex and Society – Kenneth Walker and Peter Fletcher

The psychological and social implications of various topics related to sex are here discussed in the belief that human sexuality is more than an autonomous function and involves the whole personality. (A 332) *2s 6d*

*Not for sale in the U.S.A.

W. M. Smart

THE ORIGIN OF THE EARTH

A339

This mortal-man's-eye view of the universe is given by an astronomer. It began in talks to groups of soldiers during the last war, and although it has grown a good deal since then it still addresses itself to those who are not astronomers. It refers to the latest knowledge about the cosmic and the atomic aspects of creation; it distinguishes between scientific knowledge and intuitive knowledge, and makes clear the author's authoritative grasp of the one, and his inability to dismiss the other. It propounds no personal philosophical speculations; it is modest as well as learned.

'In plain, straightforward, uncoloured English, its sober lucidity makes it as convincing as it is informative. Not least of its merits is its scientific caution: throughout it explains the evidence on which conclusions are drawn, and distinguishes reasonably acceptable theory from speculation and guesswork. . . . It is in exemplary contrast to one or two quasi-scientific romances that have recently appeared. It is popular in the best sense, and is wholly to be welcomed.' – *The Literary Guide*

2s 6d

E. T. Bell

MEN OF MATHEMATICS

A276, A277

Although their work is indispensable for an appreciation of much of science and philosophy, the great mathematicians are far less well-known than the great scientists and philosophers. Yet several of them have had interesting lives in military affairs, statecraft, and other practical pursuits. On the personal side, mathematicians have been all sorts and conditions of men, poor and rich, honest and dishonest, open-handed and close-fisted, peaceable and quarrelsome, arrogant and humble – in fact, almost anything except stupid. Like some of the famous poets and musicians, most of the creative mathematicians have matured early; some accomplished lasting work before they were twenty. The following chapters present a fair sample of the lives of these extraordinarily gifted men. Volume I includes: Zeno, Eudoxus, Archimedes, Descartes, Fermat, Pascal, Newton, Leibniz, the Bernoullis, Euler, Lagrange, La Place, Monge, Fourier, Poncelet, Gauss, and Cauchy. Volume II continues the survey up to Poincaré and Cantor, who were famous figures in the early part of the twentieth century.

2 *volumes*, 2*s* 6*d each*

Not for sale in the U.S.A. or Canada

F. M. Burnet

VIRUSES AND MAN

A265

In civilized countries most of the serious infectious diseases have disappeared or become relatively unimportant. The great majority of the infections which the average individual is likely to encounter nowadays are virus diseases – mostly nuisances rather than positive dangers – but presenting some extremely interesting problems for the epidemiologist and laboratory worker. Viruses as the smallest of living organisms are of the greatest interest to the biologist concerned with the fundamental nature of life, and the new techniques which have been developed in the last twenty years have made their laboratory study a fascinating pursuit. In this book the author gives an account of the common virus diseases of man in which both the human and the scientific aspects are considered. The theme running through almost every chapter is that of survival – how has the mumps virus survived unchanged since the days of Hippocrates? – where do influenza viruses go in summer? – and in the final chapter an attempt is made to assess what the future may hold in store. Is another influenza pandemic like that of 1918 still possible; is it conceivable that new virus diseases could be deliberately created; why has poliomyelitis become important only within the last fifty years? There are still plenty of questions.

2s

M. J. Moroney

FACTS FROM FIGURES

A236

The enormous success and rapid expansion of statistical techniques in recent years is ample proof of the need for them. They are not a cure-all, but many a headache persists because the research worker, production inspector, or executive imagines them as being too mathematical for him to apply. But there is nothing magical or mysterious in them. Statistical tools have been developed by practical men to deal with practical problems as simply as possible. Common sense and simple arithmetic will carry the reader through this book. Every symbol, every principle is explained and illustrated with examples drawn from a wide variety of subjects. The reader will find here a comprehensive introduction to the possibilities of the subject; he is given the how and the why and the wherefore by which he can recognize the kind of problem where Statistics pays dividends. The author writes from experience, for he knows the limitations to the usefulness of statistical technique, and appreciates the difficulties of the non-mathematician. The book ranges from purely descriptive statistics, through probability theory, the game of Crown and Anchor, the design of sampling schemes, production quality control, correlation and ranking methods to the analysis of variance and covariance.

3s 6d

Also by W. W. Sawyer

MATHEMATICIAN'S DELIGHT

A 121

This is the sixth edition of a volume specially written for the Pelican series in 1943. It is designed to convince the general reader that mathematics is not a forbidding science but an attractive mental exercise. Its success in this intention is confirmed by some of the reviews it evoked on its first appearance:

'It may be recommended with confidence for the light it throws upon the discovery and application of many common mathematical operations.' – *The Times Literary Supplement*

'It jumps to life from the start, and sets the reader off with his mind working intelligently and with interest. It relates mathematics to life and thought and points out the value of the practical approach by reminding us that the Pyramids were built on Euclid's principles three thousand years before Euclid thought of them.' – *John o' London's Weekly*

'The writer clearly not only loves his subject but has unusual gifts as a teacher . . . from start to finish the reader, whose own interests and training may lie in very different fields, can follow the thread.' – *Financial News*

2*s*

Lightning Source UK Ltd.
Milton Keynes UK
UKHW020652150223
417035UK00008B/127